PRAI**S**
NEUTRINO

T0058022

A *Los Angeles Times* Holiday Gift Guide Selection

A "Book to Watch Out For,"
The New Yorker's Page-Turner Blog

One of the Best Books of 2013 for the Physics Fan,
Scientific American's Cocktail Party Physics Blog

"A bow to Jayawardhana's skill in explaining how we've gotten
as far as we have in understanding [neutrinos]."
—Laurence A. Marschall, *Natural History*

"Common yet coy, neutrinos are a mystery. But they are important . . . [*Neutrino Hunters* is] comprehensive on the potential
use of neutrinos in examining the innards of the sun, of distant
exploding stars or of Earth, as well as more practical uses such
as fingering illicit nuclear-enrichment programmes."
—*The Economist*

"An intriguing story, deftly told . . . with commendable brevity
and clarity . . . Comprehensive without being overburdened with
detail or weighed down with too much theory . . . The book's
neat pen portraits of the men and women who tracked down
the poltergeist particle give it added depth. Think of this as a
great ghost story and a thumping good piece of science writing
rolled into one."   —Robin McKie, *The Observer* (London)

"Jayawardhana goes a step further than just tracking the scientific progress made in the neutrino search: he tells a story . . . Paced perfectly, with some very in-depth topics covered in a compelling and easily understandable way . . . This is a well-written and gripping history."

—Nicky Guttridge, *Sky at Night*

"[Jayawardhana] approachably and accessibly explains the search for the 'pathologically shy' elementary particles that zip through the universe without so much as an electric charge."

—*Harvard Magazine*

"A tale of revolutionary science and of the colorful personalities of those who did it—must-reading for armchair physicists!"

—Bryce Christensen, *Booklist*

"A fascinating account of the disputes between the theorists and experimentalists in this epic scientific adventure story."

—*Kirkus Reviews*

"With clarity and wry humor, Jayawardhana relates how Wolfgang Pauli 'invented' the neutrino to explain where missing energy went during beta decay . . . From deep underground in South Dakota's Homestake Gold Mine to Antarctica's IceCube, currently the world's largest neutrino detector, Jayawardhana vividly illuminates both the particle that has 'baffled and surprised' scientists, and the researchers who hunt it."

—*Publishers Weekly*

"Everything about neutrinos is fascinating. The various dramas associated with their discovery, our efforts to understand their very weird properties, and what they have taught us about fundamental physics are remarkable. Jayawardhana is the perfect person to convey these exciting stories, and *Neutrino Hunters* should be of broad interest."

—Lawrence M. Krauss, theoretical physicist and bestselling author of *The Physics of Star Trek* and *A Universe from Nothing*

"With his typical blend of scientific insight and storytelling verve, Jayawardhana vividly, colorfully, and humorously captures the often offbeat characters who, over the past century, have pursued one of the most elusive—and significant—mysteries in the history of physics."

—Richard Panek, author of *The 4% Universe*

"From the Earth's core to exploding stars, vanishing scientists, and the very essence of matter in the universe, *Neutrino Hunters* is a wild and immensely satisfying ride."

—Caleb Scharf, author of *The Copernicus Complex*

"Move over, Neil deGrasse Tyson and Brian Greene! Ray Jayawardhana is the new dean of popular science. *Neutrino Hunters* is a wonderful read from start to finish."

—Robert J. Sawyer, Hugo Award–winning author of *Red Planet Blues*

"This is science from the front lines. But it's not *just* science: *Neutrino Hunters* also illuminates the thinkers and tinkerers who have made the quest for the neutrino their life's work. We are lucky to have Jayawardhana—a first-rate storyteller who also knows the physics inside out—to guide us through this remarkable story."   —Dan Falk, author of *In Search of Time*

"*Neutrino Hunters* is an excellent overview of a vibrant and vital area of research."
—Lee Billings, author of *Five Billion Years of Solitude*

"Jayawardhana tells a whopping good ghost story. In recent years, researchers have discovered that neutrino particles, the poltergeists of physics that go right through us with nary a bump, promise to reveal much about the Earth, stars, and our cosmic origins. Beautifully written, *Neutrino Hunters* paints a vivid portrait of this new astronomy for the twenty-first century and the fascinating scientists who put it into place."
—Marcia Bartusiak, author of
The Day We Found the Universe

"*Neutrino Hunters* is a fascinating, comprehensive look at the monumental efforts to detect the least understood particle known to physics. While the Higgs boson might be more famous, Jayawardhana reveals that neutrinos are far more mysterious, and that they may hold the key to the next break-throughs in the field." —Chad Orzel, author of
How to Teach Physics to Your Dog

"*Neutrino Hunters* is a riveting mix of science and biography, providing both entertainment and painlessly assimilated information. Jayawardhana makes clear that the story is just beginning, as neutrino astronomy is starting to provide new insights into the nature of the universe. Fascinating."
—John Gribbin, author of *In Search of Schrödinger's Cat*

RAY JAYAWARDHANA
NEUTRINO HUNTERS

Ray Jayawardhana is the dean of science and a professor of physics and astronomy at York University in Toronto. His discoveries have been featured in *Newsweek*, *The Washington Post*, *The New York Times*, *The Globe and Mail* (Toronto), and *The Sydney Morning Herald*, and on the BBC, NPR, and CBC, and have led to accolades such as a Guggenheim Fellowship, the Steacie Prize, and the Rutherford Medal. He is an award-winning writer whose articles have appeared in *The Economist*, *The New York Times*, *Scientific American*, *Astronomy*, and elsewhere. He is the author of *Strange New Worlds*. Follow him on Twitter at @DrRayJay and on the Web at www.rayjay.net.

*Strange New Worlds: The Search for Alien Planets
and Life Beyond Our Solar System*

Star Factories: The Birth of Stars and Planets

NEUTRINO HUNTERS

THE THRILLING CHASE FOR
A GHOSTLY PARTICLE TO UNLOCK
THE SECRETS OF THE UNIVERSE

RAY JAYAWARDHANA

SCIENTIFIC AMERICAN / FARRAR, STRAUS AND GIROUX

NEW YORK

For my mother, Sirima, with love and gratitude

Scientific American / Farrar, Straus and Giroux
18 West 18th Street, New York 10011

Published in 2013 by Scientific American / Farrar, Straus and Giroux
First paperback edition, 2015

The Library of Congress has cataloged the hardcover edition as follows:
Jayawardhana, Ray.
    Neutrino hunters : the thrilling chase for a ghostly particle to
unlock the secrets of the universe / Ray Jayawardhana. —
First edition.
        pages   cm
    Includes bibliographical references (p.   ) and index.
    ISBN 978-0-374-22063-1 (hardback)
    1. Neutrino astrophysics.   2. Neutrinos.   3. Neutrino
interactions.   I. Title.

QB464.2 .J38 2013
523.01'97215—dc23

                                        2013021506

Paperback ISBN: 978-0-374-53521-6

Designed by Jonathan D. Lippincott

Scientific American / Farrar, Straus and Giroux books may be purchased
for educational, business, or promotional use. For information on bulk
purchases, please contact the Macmillan Corporate and Premium Sales
Department at 1-800-221-7945, extension 5442, or write to
specialmarkets@macmillan.com.

www.fsgbooks.com • books.scientificamerican.com
www.twitter.com/fsgbooks • www.facebook.com/fsgbooks

Scientific American is a trademark of Scientific American, Inc.
Used with permission.

The most exciting phrase to hear in science, the one that heralds new discoveries, is not "Eureka!" but "That's funny . . ."

—Isaac Asimov

CONTENTS

NEUTRINO HUNTERS

ONE

THE HUNT HEATS UP

There he stood, wearing a red parka, Norwegian Prime Minister Jens Stoltenberg, on blindingly white snow against a clear blue sky, 9,000 feet above sea level, with the temperature hovering at minus 20 degrees Fahrenheit. "We are here today to celebrate one of the most outstanding achievements of mankind," he bellowed out, as the sounds of flags flapping in the wind and snow crushing under a walker's boots threatened to muffle his voice. His brief remarks over, with a couple of hundred workers, guests, and tourists watching, Stoltenberg unveiled a bust carved in ice, placed atop a waist-high column: "That's the man!"

The ice sculpture bore the likeness of Stoltenberg's legendary countryman Roald Amundsen. The low-key ceremony at the bottom of the world marked the centenary of Amundsen and four mates arriving at the South Pole on December 14, 1911, delivering historic glory to the young nation of Norway, which had become independent from Sweden a mere six years earlier. Fueled by relentless determination and aided by dogsleds, Amundsen's team famously beat the ill-fated expedition led by the British naval officer Robert Falcon Scott by nearly five weeks, scoring what was undoubtedly a remarkable feat of terrestrial exploration.

Today the frozen wasteland where the fierce competition between Amundsen and Scott played out, with the pride of nations and the lives of heroes at stake, is a hotbed of activity for a different breed of explorers with more ethereal goals. Intrepid bands of scientists racing to unravel mysteries of life, our planet, and the universe are the ones laying claim to Antarctica now. In fact, the continent crawls with well over a thousand scientists and support personnel during the summer months. Geologists dig up ice cores and track the movements of glaciers for clues about climate change. Atmospheric scientists fly helium-filled balloons to measure stratospheric ozone, to complement the observations of satellites staring down from space. Paleontologists forage for fossils of creatures that were wiped out by the deadliest of known extinctions 250 million years ago. Biologists scour the dry valleys of Antarctica in search of organisms that thrive in extreme habitats. In early 2012, after many years of drilling, Russian researchers pierced through two miles of ice to reach Lake Vostok, a pristine subglacial reservoir shielded from sunlight and the wind for some 20 million years; they had hopes of encountering hitherto unknown life-forms.

Two years earlier, I got to experience what it was like to live and work on the ice when I went to Antarctica as a member of a meteorite-collecting expedition. We reached McMurdo Station, the American research center on the coast located near Scott's 1902 landing site, by military transport plane from New Zealand. After a week of preparations, packing, and training, we then flew to a seasonal base camp, where, two by two, we boarded a Twin Otter plane on skis for the final leg of our journey. The small aircraft, operated by Canadian bush pilots, dropped us off on a remote ice field just five degrees from the Pole. That's where eight of us—two women and six men—camped out in yellow, pyramid-shaped "Scott tents" for the next five bone-chilling weeks, cut off from the rest of the world except for a satellite telephone and the occasional drop-off of mail and supplies. This being the Antarctic summer, the Sun was always up, tracing a counterclockwise circle in the sky every twenty-four hours. There was no sign of life—human, animal, or plant—to be seen anywhere.

Day after day, if the winds were bearable, we went out on snowmobiles or on foot to search the nearby vast ice field and the moraines next to the hills for rocks that had fallen from space. Wrapped in big red parkas as well as thermal layers, bunny boots, neck warmers, gloves, goggles, balaclavas, and hats, we took care to avoid frostbite and crevasses during our excursions. It was easy to slip and fall on the rock-hard ice and hurt yourself badly. I slid off the Ski-Doo once, but thankfully the thick parka cushioned my fall. Others on the team also had minor mishaps, but we survived the cold, the tedium, and the isolation without any serious problems. In fact, we enjoyed the stark beauty of the landscape—the views from the tops of rocky peaks were especially magnificent—and found ways to entertain ourselves. By the expedition's end, our team had collected a total of 900 meteorites, which are now available to researchers from around the world for a variety of studies. Our own reward was the remarkable experience itself—and the delightful Adélie and emperor penguins we encountered near McMurdo at the end of the season. My one regret is that I didn't get to visit the South Pole, despite being so close to it.

The focus of activity at the Pole itself is decidedly extraterrestrial. These scientists seem to have taken to heart Marcel Proust's adage that "the only true voyage of discovery . . . would not be to travel to new lands, but to possess other eyes." The most striking part of their apparatus near the Pole is a 10-meter (33-foot) radio dish turned skyward, to map the feeble afterglow of the big bang. One of my Toronto colleagues, Keith Vander-linde, spent most of the year 2008 taking care of this telescope; he survived the polar "night" that lasted for six months, temperatures that dipped to minus 100 degrees Fahrenheit, and the overwhelming sense of isolation, not to mention the short showers and the severe boredom. But the most ambitious, and unconventional, of the scientific instruments near the South Pole is buried permanently deep under the ice, and it looks down, not up. Its construction—or burial, to be more accurate—was completed just a year before the Amundsen centennial celebration. All that the visiting dignitaries could see aboveground was a rectangular office trailer on stilts, filled with cables and computers. There was little sign of what lay beneath but for the small flags that scientists had planted helpfully on the ice to mark its mammoth footprint.

IceCube is an observatory like no other. The glacial ice itself, transparent and cleared of air bubbles by extreme pressure at depths greater than a mile, serves the same purpose as the smooth primary mirror of a conventional astronomical telescope. Buried in it are 86 long steel cables standing vertically, with 60 basketball-size globes hanging on each at regular intervals. Every one of the 5,160 globes contains optical sensors and electronics. The sensors, called phototubes, act like lightbulbs in reverse: they collect light and generate electric signals. In the case of IceCube, these sensors scrutinize the subterranean ice for faint blue flashes that occasionally shimmer in the dark stillness. Whenever a sensor detects a flash, it sends a signal to computers on the surface.

The blue flickers mark the passage of elementary particles known as muons, which belong to the same family as electrons but are about two hundred times more massive. By combining signals from the different nodes of this deeply buried sensor network, physicists can trace a muon's path in 3-D. But the researchers are not after the muons themselves. They are hunting for neutrinos, by far the most elusive and the weirdest of all known denizens of the subatomic world. These ghostly particles interact every once in a while with protons within ice molecules to release muons, thus betraying their presence as the muons in turn light up the ice. Since a newly created muon travels through ice along the same path as the incoming neutrino did, researchers can tell which direction the neutrino came from by examining the muon's trail.

Neutrinos are elementary particles, just like electrons that buzz about atomic nuclei or quarks that combine to make protons and neutrons. They are fundamental building blocks of matter, but they don't remain trapped inside atoms. Also unlike their subatomic cousins, neutrinos carry no electric charge, have a tiny mass, and hardly ever interact with other particles. A typical neutrino can travel through a light-year's worth of lead without interacting with any atoms. Therein lies the snag: neutrinos are pathologically shy. Their severe reluctance to mingle makes these particles hard to pin down, so neutrino hunting is a tricky business. But every so often, a neutrino does collide with something, such as a proton inside a water molecule, essentially by accident. It is to raise the odds of accidental collisions, and thus to increase our chances of observing neutrinos, that scientists build extremely large detectors like IceCube.

You still can't see neutrinos directly, but you can get a whiff of their presence from the clues they leave behind. On the rare occasions that neutrinos *do* interact with matter, they produce charged particles such as muons that physicists can detect with their instruments. But distinguishing neutrino signals from

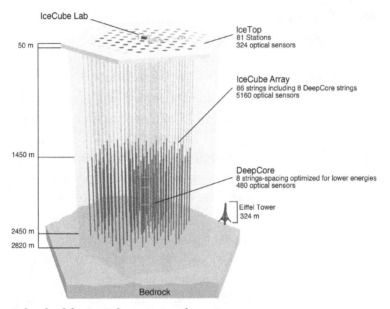

IceCube Lab

IceTop
81 Stations
324 optical sensors

50 m

IceCube Array
86 strings including 8 DeepCore strings
5160 optical sensors

1450 m

DeepCore
8 strings-spacing optimized for lower energies
480 optical sensors

Eiffel Tower
324 m

2450 m

2820 m

Bedrock

A sketch of the IceCube neutrino observatory (J. Yang/NSF)

unrelated "noise" poses a challenge: cosmic rays, fast-moving particles that arrive from deep space, also produce muons, which might be confused with muons produced by neutrino interactions. Neutrino hunters place their equipment deep underground, or under a thick layer of ice, so that cosmic ray muons cannot get through. As Janet Conrad of the Massachusetts Institute of Technology explains, "If you're trying to listen to a whisper, you don't want a lot of noise around."

Neutrinos are hard to catch, but they are also among the "most wanted" of all cosmic messengers for the secrets they hold about the nature of matter, the pyrotechnics of exploding stars, and the structure of the universe itself. Besides, in the words of theorist Boris Kayser of the Fermi National Accelerator Laboratory (Fermilab) near Chicago, which is home to several neutrino experiments, "If neutrinos didn't exist, we wouldn't be here." He explains that "the Sun produces energy through nuclear reactions on which life on Earth depends, and those reactions could not occur without neutrinos." Moreover, the nuclear burning in previous generations of stars, which produced the heavy elements necessary for life, would not have been possible without neutrinos, either. Therefore, he argues that "to make sense of the universe we need to understand neutrinos well."

Thankfully, neutrinos are as ubiquitous as they are cagey. In fact, neutrinos are the most abundant matter particles in the universe. According to Hitoshi Murayama of the University of Tokyo and the University of California, Berkeley, there are a billion neutrinos for every atom in the universe. He contends that "their sheer number means they have an important role. The contribution of neutrinos to the cosmic energy budget is comparable to that of all the stars." In fact, about a hundred trillion neutrinos produced in the nuclear furnace at the Sun's core pass through your body every second of the day *and* night, yet they do no harm and leave no trace. During your entire lifetime, perhaps one single neutrino would interact with an atom in your body. Neutrinos travel right through the Earth unhindered, like bullets cutting through fog. Besides, the Earth's bowels generate neutrinos, as radioactive elements decay, and so do collisions of energetic particles from space in the upper levels of the atmosphere. Cataclysmic deaths of massive stars set off tremendous bursts of neutrinos, which escape these sites of mayhem unscathed and bring us news of awesome celestial events millions of light-years away. Moreover, our planet is immersed in a sea of cosmic neutrinos, which sprang forth when the infant universe was barely two seconds old.

The bizarre traits of neutrinos have turned them into pop culture icons of sorts. As far back as 1960, John Updike celebrated them in a delightful poem published in *The New Yorker*. Titled "Cosmic Gall," it described how neutrinos traverse the Earth as easily as dust bunnies travel down a drafty hall or light passes through a sheet of glass. Klaatu, a Canadian progressive rock band perhaps best remembered for false rumors that they were the Beatles recording under a pseudonym, described the same phantom behavior, of neutrinos passing right through our bodies without alerting us, in the lyrics of a 1976 song. Neutrinos have even starred as hipster characters in the animated television series *Teenage Mutant Ninja Turtles*.

Not surprisingly, references to neutrinos have also popped up on the popular sitcom *The Big Bang Theory*, in which two of the main characters are physicists. The show's science consultant, David Saltzberg of the University of California, Los Angeles, is himself a physicist who works on neutrino telescopes, among other topics. In one scene, the co-lead Sheldon Cooper is fiddling with equations on a whiteboard in his office when his fellow physicist and roommate Leonard Hofstadter enters along with their engineer friend Howard Wolowitz. Sheldon exclaims, "Oh, there's my missing neutrino. You were hiding from me as an unbalanced charge, weren't you, you little subatomic Dickens?" Instead of acknowledging his friend's greeting, he continues, "Here, look, look, I found my missing neutrino." Howard responds drily, "Oh, good, we can take it off the milk cartons."

Neutrinos have made numerous appearances in science fiction, of course, typically as the culprits behind strange or catastrophic events. In Robert J. Sawyer's novel *Flashforward*, a burst of neutrinos from a dying star is responsible for making everyone lose consciousness briefly and see themselves as they would be some twenty-one years in the future. In Greg Bear's

Foundation and Chaos, a freak neutrino storm wipes out the rules that robots are programmed to follow (à la Isaac Asimov's original Foundation series), resulting in complete mayhem. More recently, neutrinos were blamed for heating the Earth's core, triggering ferocious earthquakes and floods, in the Hollywood disaster flick *2012* directed by Roland Emmerich.

Despite neutrinos' quirky appeal as cultural icons, few people outside the physics community paid much attention to the science of real-life neutrinos until they made headlines recently for possibly breaking the cosmic speed limit set by Albert Einstein back in 1905. A large international collaboration of physicists known as OPERA (acronym for the unwieldy title Oscillation Project with Emulsion-tRacking Apparatus) made the startling announcement in a research paper posted online and at a press conference in late September of 2011. The particles appeared to travel faster than light between CERN, the European Organization for Nuclear Research and its Laboratory for Particle Physics in Geneva, Switzerland, and an underground detector 454 miles away in Gran Sasso, Italy, arriving 60 nanoseconds sooner than expected.

Despite the OPERA spokesman's cautionary words, and skepticism from the vast majority of neutrino researchers, the news reverberated around the globe. Perhaps the commotion was not surprising given the astounding implications. If true, the finding would violate Einstein's theory of special relativity, a cornerstone of modern physics. As *Time* magazine put it, "If the Europeans are right, Einstein was not just wrong but almost clueless." Most physicists and journalists emphasized that the extraordinary claim required further investigation and independent verification. "If true, it is a result that would change the world. But that 'if' is enormous," said *The New York Times*.

But all that hedging failed to rein in rampant speculations

about superluminal voyages and grandiose visions of new physics. Suddenly jokes about neutrino time travel were everywhere. Some quipped that neutrinos had obeyed the law in Switzerland but broken the speed limit once they crossed over to Italy. On the sitcom *The Big Bang Theory*, Sheldon Cooper tried to foster dinnertime conversation by asking, "Faster-than-light particles at CERN: paradigm-shifting discovery or another Swiss export as full of holes as their cheese?" The Irish folk band Corrigan Brothers, which had performed at President Barack Obama's 2009 inauguration, posted a song they performed with Pete Creighton on YouTube, questioning whether $E=mc^2$ still held true now that neutrinos appeared to travel faster than light. Toward the end, however, the song lyrics cautioned against rushing to conclusions and suggested that Einstein might still be right about the cosmic speed limit.

As a way of accommodating the new result without breaking the light barrier, some theorists proposed that the Swiss neutrinos might have tunneled through a hidden extra dimension, *Star Trek*–style, on their way to Italy, reducing the distance they needed to travel. Others suggested a different shortcut due to a crumpling of space-time near the Earth. Many critics pointed to possible experimental errors. Andrew Cohen and Sheldon Glashow of Boston University raised a serious theoretical gripe: a beam of superluminal neutrinos would rapidly lose energy by emitting other particles, so the beam should be depleted of high-energy neutrinos by the time it reached Gran Sasso, something that was not observed. Meanwhile, a second, more precise speed test done by the same OPERA collaboration, announced in mid-November, bolstered the surprising result.

Three months later CERN released a brief but crucial update. It read, in part: "The OPERA collaboration has informed

its funding agencies and host laboratories that it has identified
two possible effects that could have an influence on its neutrino
timing measurement . . . If confirmed, one would increase the
size of the measured effect, the other would diminish it." The
first effect, a possible problem with time stamps of GPS units
used to synchronize the clocks at the two sites, would actually
make the neutrino speed even faster than previously reported.
The other effect, a bad cable connection between a GPS unit
and a computer, would mean that the neutrinos had in fact
traveled slower than light. Most media reports and commenta-
tors focused on the latter possibility. *The Wall Street Journal*
described it as "a potentially embarrassing outcome" for the researchers involved. Then, on March 16, 2012, a different team
of physicists, whose ICARUS detector is also located at Gran
Sasso, reported a new measurement of flight time for neutrinos
from CERN: their speed did not exceed that of light. "The evidence is beginning to point towards the OPERA result being
an artifact of the measurement," said CERN's research director, Sergio Bertolucci.

Even though neutrinos turned out not to be superluminal
in the end, they have taught us a great deal already about the
shenanigans of the subatomic realm and allowed us to peer
deep into the Sun's scorching heart. Besides, without neutrinos,
nuclear power generators and nuclear bombs would not be possible. Neutrinos were the first harbingers of the dramatic demise of a massive, bloated star that exploded 160,000 light-years
away in the Large Magellanic Cloud, a satellite galaxy of the
Milky Way that appears as a fuzzy patch in the southern sky.
Three underground detectors in Japan, Russia, and the United
States recorded a total of two dozen neutrinos from the explosion, out of the billions upon billions that swept through the

Earth, in a short burst on February 23, 1987. It was only a few hours later that astronomers scanning the skies from a far-flung mountaintop observatory in Chile saw the supernova in visible light.

Over the years, neutrinos have drawn the attention of some of the most brilliant minds and colorful personalities in the history of physics. The cast of historical characters associated with neutrinos included the sharp-witted Wolfgang Pauli, who invoked these particles in the first place to dodge a crisis in physics; the troubled genius Ettore Majorana, who theorized about neutrinos' mirror twins before disappearing without a trace at the age of thirty-two; and the committed socialist Bruno Pontecorvo, who realized that neutrinos might morph between different types and caused a Cold War ruckus by defecting to the Soviet Union. Some neutrino hunters built experiments deep underground to peer into the heart of the Sun, while others set up traps next to powerful nuclear reactors to catch neutrinos changing form. During the past two decades, many more scientists have caught the neutrino bug and joined the quest.

That's because, for neutrino hunters, the best is yet to come. These shadowy particles promise to unlock some of the greatest secrets of the universe. They could tell us about the birth sites of enigmatic cosmic rays that bombard the Earth around the clock. For astronomers, who have had to rely almost exclusively on electromagnetic radiation in the form of visible light, radio waves, and X-rays from distant celestial bodies, neutrinos offer an exciting new window on the most violent phenomena in nature. In fact, neutrinos may have a lot to do with triggering spectacular stellar explosions in the first place. Some scientists have proposed that a sterile variety of neutrinos could account for so-called dark matter, which makes up nearly a quarter of the universe but remains undetected except through its gravitational tug on galaxies. The imprints left by primordial neutrinos on the faint afterglow of the big bang, which is still measurable with microwave telescopes, could reveal the conditions very soon after the universe was born.

What's more, we may have neutrinos to thank for the simple fact that the universe is not empty of matter, and thus for our very existence. Just after the big bang, there was lots of energy to give rise to particle and antiparticle pairs. The cosmic density was so high back then that these pairs should have come together and annihilated each other quickly, leaving only a sea of radiation. To avoid the catastrophe, there must have been a tiny preponderance of normal matter over antimatter. Physicists struggle to understand how such an asymmetry came about. One popular explanation is that super-heavy cousins of neutrinos in the early universe decayed in such a way as to make one extra particle of matter for every billion particle-antiparticle pairs. Measuring subtle properties of today's light neutrinos could tell us whether such a scenario was indeed responsible for tipping the balance ever so slightly in favor of matter. As Boris Kayser points out, "Again, if not for neutrinos, we may not be here."

Most exciting, if not unsettling, is the prospect of physics beyond the so-called standard model. Formulated in the early 1970s, the standard model incorporates two dozen elementary particles of matter and their antimatter twins, three types of interactions among them, and the symmetries that govern those interactions. It is the best description of the subatomic world that we have, and countless experiments over three decades have verified its predictions with exquisite precision. The fabled Large Hadron Collider at CERN, the most powerful and expensive atom smasher ever, was constructed at a jaw-dropping price tag of roughly $9 billion in large part to nail down the final missing piece of the theory. The LHC confirmed the existence of the Higgs boson, a particle hypothesized to be responsible for endowing other elementary particles with mass. The standard model, however, presumed that neutrinos have no mass, come in three flavors, and cannot change form. So the discovery that neutrinos do have a very small but nonzero mass, and a chameleonlike tendency to morph among the three types, has exposed a crack in the model's elegant edifice. If it turns out that there are more than three neutrino flavors, as some data hint, such a revelation could shatter the very foundations of physics. As physicist Kate Scholberg of Duke University puts it, "We're right on the verge of exploring a new regime in physics. Several unknowns out there are teasing us." She points out that "neutrinos provide us with a whole new sector of phenomena that we can measure to investigate the nature of the universe."

The starring role of neutrinos in a great many sagas unfolding across physics, cosmology, and astronomy explains why scientists make considerable efforts to trap these minuscule particles. Over the past two decades, they have built ever more sophisticated neutrino experiments dotting the globe. From a deep nickel mine in Ontario to a freeway tunnel crossing a mountain in central Italy, and from a nuclear waste site in New Mexico to a bay on the South China Sea, neutrino hunters are chasing their quarry.

The most impressive of their traps remains IceCube, the world's biggest neutrino telescope, built at a cost of over $270 million. Its completion is a long-held dream come true for its visionary director, Francis Halzen. Growing up in Belgium,

Halzen hoped to become a schoolteacher, but at university he got interested in physics, and he never looked back. After working at CERN for a few years, he moved to the University of Wisconsin–Madison, where he has been a professor for four decades. As a theoretical physicist, he worked on some aspects of quantum mechanics before he turned his attention to neutrino hunting in the mid-1980s. Halzen first heard about attempts to detect neutrinos in Antarctica from colleagues at the University of Kansas, while he was visiting there to deliver a lecture. They told him that Russian scientists had been using radio antennas at their Antarctic research station to search for electric sparks resulting from cosmic neutrinos colliding with the ice. Halzen found the experiment intriguing, and together with two colleagues, he set out to calculate how strong such signals would be. They were disappointed to find that radio emissions produced by most neutrino interactions would be far too weak to register. They concluded that the Russian experiment was doomed to fail. Instead, they realized, it would make more sense to look for bursts of blue light in the ice, which would also indicate the arrival of neutrinos. Halzen was convinced that sinking an array of light sensors deep into the Antarctic ice was a great way to catch neutrinos coming from the far reaches of space.

Excited about the prospect of developing a novel neutrino telescope in Antarctica, Halzen e-mailed several other physicists to ask what they thought about his idea. John Learned at the University of Hawaii was among them. Born to an old New England family whose ancestors included a general who fought in the American Revolutionary War, Learned grew up on Staten Island and spent his childhood summers with his grandparents in upstate New York. He enjoyed acting like an outsider in both places: "In the country I was a city kid and in the city I was a country kid," he says. In middle school, he worked on the school newspaper and ran a weather station on the school roof. He also remembers volunteering to take a traveling science exhibit around the school "because it was a great excuse to get out of the classroom." Later, at Brooklyn Technical High School, he took many shop classes, which have served him well as an experimental physicist. As a graduate student at the University of Washington, Learned investigated the prospects for measuring cosmic ray particles underwater. He built a barge, anchored it in the middle of Lake Chelan, and lowered particle detectors into the clear, deep water. After completing his doctorate, he took a job at a research station at Echo Lake in the Rocky Mountains of Colorado, where he lived in a log cabin with his wife and two small children. It was during that period that Learned got seriously interested in neutrinos, and he later moved to Hawaii in the hope of deploying a giant array of underwater neutrino detectors in deep waters of the Pacific Ocean surrounding the volcanic islands. Given Learned's interests and expertise, it was no surprise that Halzen reached out to him for his opinion about burying neutrino detectors in Antarctica.

The two physicists discussed the attractions of Halzen's scheme. "Learned immediately appreciated the advantages of an Antarctic neutrino telescope," according to Halzen. For starters, polar ice is clear, dark, stable, and sterile, and free of background light from bioluminescent organisms and emissions from radioactive decay of sea salt, which could confuse detection of the neutrino signals. Equally important was the fact that the National Science Foundation (NSF) was already operating a research base at the South Pole, and therefore was in a position to provide vital logistical support. Encouraged by Learned's enthusiasm and input on detector design, Halzen presented their concept at a conference in Poland, and wrote it up in a paper in 1987, but left it at that, perhaps because as a theorist he didn't have experience building experiments and hesitated to take on such an ambitious task himself.

Halzen recalls a phone call he received from an irate official at the NSF about a year later. The official complained that two young physicists from the University of California, Berkeley, had tried to sneak a string of phototubes into Antarctica to place inside a drill hole, without proper authorization. He asked whether Halzen was responsible for putting "this crazy idea" into their heads. Halzen assured the official that he had never heard of the two Berkeley physicists, who had apparently gotten the notion from attending a conference where Halzen and Learned had discussed their scheme.

Later, Halzen teamed up with the Berkeley group to pursue the idea in earnest. First they tested its feasibility by sinking a 200-meter-long (656-foot) strand with three phototubes into a hole drilled by glaciologists in Greenland. Then they began work on a pilot experiment called AMANDA (for Antarctic Muon and Neutrino Detector Array), with funding and support from the NSF, in the austral summer of 1992. They borrowed a technology that glaciologists had developed to bore holes in the ice: a drill that shot out hot water, as if from a high-pressure showerhead, which melted its way down. The cavity did not refreeze for several days, giving them enough time to deploy the sensors attached to a cable.

As the team lowered the first string of phototubes into the ice on Christmas Eve of 1993, Halzen was at his family's house in Belgium. Being a theorist, he wasn't needed at the work site. He had a lot at stake, though, and glanced often at a computer on his lap during the meal, hoping for e-mail updates from the South Pole. As he wrote later, "To have your career on the line half a world away is hard enough. But to know that you have embroiled so many others in the same improbable adventure, that your funders and colleagues expect results, and that you are totally powerless to affect the outcome, is a form of exquisite torture." Just as dessert was being served, Halzen was relieved to receive a message confirming that the deployment was a success.

The team's delight didn't last long, because they encountered unexpected challenges. One problem was that the phototubes registered lots of blue flashes from muons created by cosmic rays. The researchers had expected the cosmic ray muons to peter out by the time they reached half a mile beneath the ice, making it easy to identify the few muons generated by neutrinos arriving from below, from the other side of the Earth. That wasn't the case: what they saw was "a nearly meaningless blur," as Halzen put it. But their biggest problem had to do with air bubbles in the ice, which scattered the blue flashes generated by neutrino events, making them harder to pinpoint. They found lots of bubbles at this depth, and the bubbles were fifty times bigger than they had anticipated. So the project was delayed while the team figured out what improvements to make. The solution, they found, was to dig *deeper* holes and sink the sensors down to a mile below the surface. At these greater depths, the researchers would have a clearer view of the blue flashes associated with neutrino arrivals, because higher pressures would squeeze out the bubbles in the ice.

The AMANDA experiment, from the first drilling to the final shutdown, lasted for a decade. (Meanwhile, Learned and his collaborators had abandoned their project off the coast of

Hawaii after many years of effort, because of technical problems.) Over that time, Halzen and his colleagues learned a lot about Antarctic ice as well as neutrino detection. Drawing on their experience, the team started construction of IceCube, designed to be a hundred times bigger than its predecessor, in 2005.

IceCube is truly a marvel of extreme engineering. Just as with AMANDA, not only its components, drilling equipment, and personnel, but also food and fuel, had to be transported to Antarctica from various parts of the world. Ski-equipped C-130 Hercules cargo planes, operated by U.S. Air Force crews, hauled them for the last leg of the journey, from McMurdo Station on the Antarctic coast to the South Pole, a distance of 800 air miles. Engineers used a custom-built high-pressure drill, with hot water shooting out of a nozzle at its end, to puncture the ice sheet to a depth of a mile and a half. It took two days of nonstop drilling, and 4,800 gallons of gasoline, to bore one hole, melting 200,000 gallons of ice in the process. Once the shaft was clear, they gently lowered the steel cable with the sensors. Hole by hole, IceCube was "built" over six austral summers, taking advantage of continuous sunlight and relatively balmy working conditions from November to February.

For Halzen, the project's completion in December 2010 was "a great relief." "Now that IceCube is built, people forget how incredibly risky and challenging this undertaking was. I've made a list of all the points when I thought the project had failed," he added. There was little room for error, with the biting cold, high altitude (of over 9,000 feet above sea level), and dreadful isolation exacerbating the risks. Once during construction a worker mistakenly grabbed a hose hanging from a drill tower, and was thrown on his back on the rock-hard ice when the hose pulled up. The victim had to be flown to New Zealand

Lowering a steel cable into the Antarctic ice (M. Krasberg/NSF)

for treatment, and it took a few weeks for him to recover completely.

Yet the gamble that Halzen and his colleagues took in building IceCube has begun to pay off already. In the first couple of years of its operation, the observatory has recorded two unusual neutrino signals, with energies far greater than any seen before. At a conference in Kyoto in the summer of 2012, team member Aya Ishihara of Chiba University in Japan showed these "PeV events," so dubbed because their energies are in the "peta-electron volt" (or quadrillion electron volt) range, corresponding to about a million times the mass-energy of a proton. The extreme energies surprised many astrophysicists. As Spencer Klein of the Lawrence Berkeley National Laboratory in California points out, "These neutrinos have energies more than

The IceCube neutrino observatory, illuminated by moonlight (E. Jacobi/NSF)

a thousand times higher than any neutrinos that we have produced in particle accelerators."

At first, the researchers wondered whether collisions between highly energetic cosmic rays and oxygen or nitrogen atoms in the Earth's atmosphere were responsible for producing these PeV neutrinos. After further monitoring and analysis, they're now convinced that's probably not the case. As Halzen puts it, "It is unlikely that they are atmospheric, and that is the exciting part." In other words, we may have to look to distant celestial sources to uncover the violent origins of these neutrinos. In fact, researchers think the particles may come from powerful jets shot out by monstrous black holes at the hearts of galaxies, or from incredible explosions known as gamma ray bursts (GRBs), which appear to be even more potent than supernovae.

Over the past two decades, astronomers have confirmed that many galaxies, including our own Milky Way, harbor gigantic black holes at their centers, and have observed high-speed jets coming off their poles. They think these jets form because black holes drag matter in from their surroundings and launch some of that material back into space with the help of magnetic fields. The particles in the jets, accelerated to speeds close to that of light, could produce energetic neutrinos, such as those that IceCube has detected. Researchers have speculated that GRBs, which might herald the death of very massive stars, could be another source of high-energy neutrinos. Discovered by chance in the late 1960s by satellites designed to look for gamma rays from secret nuclear tests in space, GRBs have confounded scientists for decades. Recent findings suggest that most GRBs consist of narrow beams of fast-moving particles ejected during the collapse of hefty stars into black holes or neutron stars. In either case, IceCube may have captured messengers coming straight from the scene of the action, so it could help us better understand some of the most ferocious phenomena in the universe.

IceCube is just the most exotic of a new generation of neutrino facilities with unprecedented sensitivity. Some, like IceCube itself and an even bigger network to be deployed on the Mediterranean seafloor, are, as we've seen, designed to catch neutrinos coming from outer space or produced when cosmic rays hit the Earth's atmosphere. Others, such as the cathedral-scale detector under Mount Kamioka in Japan and another, weighing nearly as much as 5,000 automobiles, tucked away in a Minnesota mine, measure neutrino beams generated by giant particle accelerators hundreds of miles away. Experiments of yet another type, located at the village of Chooz in France and at

Daya Bay in China, harness neutrinos produced in commercial nuclear power plants.

Together, these facilities make up the formidable arsenal of today's neutrino hunters. Their advent signals that neutrino chasing, once an esoteric sideline, is now ready for prime time. In the coming chapters, we will follow that thrilling chase, along with its bewildering twists, just as we enter a brave new era that promises to unravel cardinal mysteries of the universe, from its puniest scale to its grandest, and quite possibly upend our most cherished theories about the nature of things. Along the way, we will meet the men and women who have made it their business to track down this most elusive of particles—from the early theorists who laid the groundwork for teasing out the neutrino's existence to the modern experimentalists who try to make sense of its quirky character—and we will catch glimpses of their heroic endeavors and fascinating lives.

TWO

A TERRIBLE THING

The first third of the twentieth century was an exhilarating time for physicists. Not one, but two sweeping revolutions were under way. On the cosmic scale, the special and general theories of relativity altered our understanding of space, time, motion, and gravity in truly fundamental ways. On the subatomic scale, the newfangled theory of quantum mechanics revealed a weird world full of strange phenomena where uncertainty rules the day and paradoxes run free.

The iconic Albert Einstein proposed the special theory at the tender age of twenty-six, while working as a clerk in the Swiss patent office, in 1905. The idea behind it was not entirely new: Galileo had pointed out three centuries earlier that all motion is relative and that there is no absolute state of rest. (That is why objects on a ship sailing at a uniform speed behave the same way as they do when the ship is docked at harbor. It also explains why we are not swept off the ground as the Earth rotates.) But Einstein extended the concept dramatically with his suggestion that the passage of time is relative while the speed of light is not. According to Einstein, light travels at the same constant speed in a vacuum no matter what, even if you were to measure it while zipping along at nearly the speed of light. The math works out to make it so, but it does lead to strange phenomena, such as time passing more slowly for those traveling very fast. If a future astronaut were to fly through space at 95 percent of light speed, she would age much less than her twin staying back on Earth. As peculiar as it may seem, scientists have verified that "time dilation" is real by flying atomic clocks around the world on airplanes and comparing them to identical clocks left on the ground. They have also confirmed the phenomenon many times over by measuring the change in the lifetime of subatomic particles traveling at different speeds.

Ten years after he introduced the special theory, Einstein went on to present the general theory of relativity to describe gravity in a novel way. Two and a half centuries earlier, Isaac Newton had treated gravity as an attractive force between objects, and his approximations are still plenty good enough for most practical purposes, even for sending spacecraft to the Moon. With his general theory, however, Einstein proposed to look at gravity as geometry, a curvature in space-time caused by the presence of a mass. His majestic equations supersede Newton's law of gravity, because they can cope well with extreme speeds and potent gravitational fields such as those encountered in the vicinity of black holes. Even in the relatively weak gravitational field of the Earth, today's GPS devices would fail miserably if we didn't use Einstein's formulas: engineers program the clocks on board satellites to account for the effects of both special and general relativity so that they will run at the same rate as clocks on the surface of the Earth.

Meanwhile, over the first three decades of the twentieth century, the likes of Max Planck, Niels Bohr, and Werner Heisenberg developed quantum mechanics to describe the behavior of matter and radiation at the subatomic level. Planck and others proposed that energy comes in discrete bundles, called quanta, which atoms absorb or emit as they jump to higher or lower energy levels. This realization led physicists to a whole new way of looking at the material world. They could now describe light, which classical physics viewed as waves, in terms of particles, called photons. Conversely, they found that subatomic particles such as electrons also exhibited wavelike properties. Physicists found that this "wave-particle duality," in which particles and waves were neither one nor the other but incorporated properties of both, provided a better description of atomic structure and the interaction between radiation and matter. What's more, Heisenberg pointed out that in the quantum world there is an inherent uncertainty to how well you can pin down the properties of a particle. The determinism of classical mechanics gave way to mere statistical probabilities: instead of predicting definitive outcomes, quantum mechanics assigns likelihood to different results. That meant, as Erwin Schrödinger pointed out in a famous thought experiment, a cat in a box could be both alive and dead at the same time, until an observer intervenes by looking in. The consequences are so bizarre that Bohr is reported to have said, "Those who are not shocked when they first come across quantum theory cannot possibly have understood it."

Remarkably, during this period of major upheaval in physics, theory and experiment kept pace with each other. Sometimes experiments provided dramatic confirmation of a theoretical prediction, as when astronomers observed during a solar eclipse in 1919 that the Sun's gravity could bend the light from a distant star, just as Einstein's general theory of relativity had anticipated. On other occasions, the application of a new theory led to a satisfactory explanation of experimental data. For example, Bohr used the concept of quantized energy to account for the spectral lines produced by hydrogen atoms absorbing light: he suggested that the lines resulted from electrons jumping between fixed orbits around the atomic nucleus. Einstein himself had used the quantum hypothesis to explain how shining light on certain materials would release electrons from that material. Once in a while, however, results of experiments uncovered new contradictions nudging theorists to come up with better descriptions of nature.

It was in this heady atmosphere that the neutrino was invented, or willed into existence, in a form of scientific witchcraft to dodge a growing crisis in nuclear physics, long before the presence of such a particle was detected through experiments. When scientists couldn't account for energy that went missing during radioactive beta decay, one theorist found it necessary to "invent" a new particle to account for the missing energy. The man behind the theoretical wizardry was a brash young physicist by the name of Wolfgang Pauli.

Born in Vienna in the year 1900, Pauli grew up in a home filled with stimulating conversations. His father was a well-known chemistry professor, and his mother was a journalist who wrote theater critiques, historical essays, and political articles with a socialist bent. "Wolfi," as his family called him, spent a happy childhood playing with his younger sister, exploring the woods near their house, and swimming in the Danube River. In high school he was popular with his classmates and displayed a knack for mocking his teachers: he tagged the nickname *das U-boot*—the submarine—on a short teacher who had a tendency to show up unexpectedly amid groups of students. His high school grades in Latin and Greek were merely satisfactory, but he excelled in mathematics and physics. Soon he found the science classes too easy, so his father arranged for private lessons in advanced physics.

It was Pauli's tutor who introduced him to Einstein's general theory of relativity. Few physicists understood the elegant but radical theory or grasped its profound implications at the time. Pauli, however, had no trouble diving in. Barely two months out of high school, he wrote a paper of his own on the subject. Determined to pursue a career in physics, he moved to Munich in 1918 to study under Arnold Sommerfeld, a pioneer in the emerging field of quantum mechanics. Pauli's paper, which had even come to Einstein's attention, impressed Sommerfeld, who wrote to a colleague about it, noting, "I have around me a really astonishing specimen of the intellectual elite of Vienna in the young Pauli . . . a first-year student!"

Pauli completed a doctorate in quantum mechanics under Sommerfeld's guidance within three years. Not long after, at the request of his thesis adviser, he wrote a review article on relativity for the *Encyclopedia of Mathematical Sciences*. His 240-page "article" was also published later as a monograph. Having read Pauli's masterpiece, Einstein raved: "Whoever studies this mature and grandly conceived work would not believe that its author is a twenty-one year old man. One wonders what to admire most, the psychological understanding of the development of ideas, the sureness of mathematical deduction, the profound physical insight, the capacity for lucid, systematical presentation, the knowledge of the literature, the complete treatment of the subject matter, or the sureness of critical appraisal."

Soon Pauli was corresponding with leading physicists throughout Europe. His letters, signed jokingly "The Scourge of God," were known for wit and sarcasm as well as scathing

Albert Einstein and Wolfgang Pauli (Pauli Archive, CERN)

criticism. Colleagues reported that he uttered the phrase "Not only is it not right, it is not even wrong" to disparage theories that were seen to lack rigor and testable hypotheses. On one occasion, after Einstein gave a lecture on relativity in Berlin, while senior professors in the audience sat in silence wondering who should ask the first question, the brazen Pauli got up and announced, "What Professor Einstein has just said is not really as stupid as it may have sounded." On another occasion, he made so many critical remarks about a lecture given by Paul Ehrenfest, a Dutch physicist twenty years his senior, that Ehrenfest told him, "I think I like your publications better than I like you." Pauli snapped back, "That's strange. My feeling about you is just the opposite." The two became friends, and continued to try to one-up each other's quips. Pauli's outspoken manner did not endear him to everyone, but he earned the respect of many of his colleagues not just for his brilliance but also for his honesty and forthrightness. Many of them saw him as the "conscience of physics," and often asked, "What does Pauli think?" when they were presented with a new idea.

After extended visits to Göttingen and Copenhagen to work with prominent physicists, Pauli took up a research assistant position in Hamburg. While there, at the age of twenty-five, he formulated what is now known as the "Pauli exclusion principle" in quantum mechanics, a key insight for understanding not only the behavior of a class of subatomic particles known as fermions (which includes electrons, protons, and neutrons) but also the inner workings of stars. The principle states that no two fermions can occupy the same "quantum state" at the same time, meaning they cannot have the same spin and energy. At the subatomic scale, it accounts for the shell structure inside atoms: at most two electrons, with opposite spin, can reside at the same energy level, so any additional electrons must occupy higher levels. At the cosmic scale, it is the reason why white dwarfs, the compact stellar cinders left over after stars like the Sun run out of fuel, hold up without shrinking further. The matter inside white dwarfs is so compressed that electrons are packed together as tightly as they can be with no more than two particles occupying the same energy level. Once that happens, gravity cannot shrink the white dwarf any farther (at least not without adding a whole lot more mass), so it resists collapsing into a black hole. Pauli's discovery of the exclusion principle earned him a Nobel Prize two decades later, because of its importance for making sense of a wide variety of physical phenomena.

His late twenties and early thirties turned out to be a tumultuous period for Pauli. His father, who had long been a womanizer, left his mother for a much younger woman whom Pauli detested. Not long after her husband left, Pauli's mother committed suicide by taking poison. His father's betrayal and his mother's tragic death traumatized Pauli. Yet the same month that his mother died, Pauli also received some good news: the Swiss Federal Institute of Technology offered him a professorship, despite his reputation for being a poor lecturer. He arrived in Zurich in April 1928, "dressed like a tourist with a rucksack on my back," according to his own description, to take up the prestigious position.

The change of scenery did little to raise Pauli's spirits, however. He had trouble making progress in his research, and considered giving up physics and writing a utopian novel instead. Frustrated, he wrote to Bohr that the problem was not a lack of time for science: "I am only stupid and lazy. I think that somebody should give me a thrashing every day! But since, unfortunately, nobody is doing this, I must look for other means to reinvigorate my interest in physics." Perhaps Pauli was distracted by the pleasures of life in Zurich: he swam in the lake, dined at elegant restaurants, hung out at beer gardens with his colleagues, and socialized with prominent lawyers, writers, and artists. After a hiatus of several months, he did manage to complete two important papers with Heisenberg on the theory of quantum electrodynamics, which describes how light and matter interact.

It was around this time that Pauli ran into a cabaret dancer by the name of Käthe Deppner at a friend's party. He had met Deppner some years earlier in Berlin. On this occasion the two developed an instant rapport, and they got married in a hurry in December 1929. But their alliance was doomed from the start. She had fallen in love with another man even before the marriage took place, and refused to give up that relationship. Pauli knew about her affair, and struggled to cope. He joked about being married in "a loose way" and about sending a printed notice to his friends in case his wife was to run away with her lover. Less than a year into the marriage, their ill-fated union ended in divorce. Crushed and bitter, Pauli bemoaned that she had left him for a mediocre chemist. "If it had been a bullfighter—with someone like that I could not have competed—but such an average chemist!" he lamented.

Somehow, despite all this personal turmoil, Pauli was able to revive his scientific focus and creativity. Foremost among his intellectual concerns at the time was a brewing calamity in the world of nuclear physics. By 1930, the theory of quantum mechanics reigned triumphant, with many stunning successes to its credit. But a pesky problem persisted: physicists noticed that when a radioactive atom spat out an electron every so often, some of its energy seemed to vanish. That was at odds with the sacrosanct physical law of energy conservation: what goes in must equal what comes out. Many of the world's top physicists, including Pauli, were troubled by the discrepancy, as it pointed to a fundamental flaw in their understanding.

The saga actually began with the serendipitous discovery of radioactivity by the French physicist Henri Becquerel back in 1896. Becquerel found that photographic plates he had left in a drawer with some uranium salts for several days had smudges on them, as if they had been exposed to light. He reckoned that the salts had emitted some form of radiation, and through a series of follow-up investigations, he showed that this radiation was an intrinsic property of the element uranium. Becquerel's curious finding intrigued a number of scientists. At Cambridge University, J. J. Thomson, best known for his discovery of the electron, steered the attention of his graduate student Ernest Rutherford to the new discovery. Born in rural New Zealand as the fourth of twelve children to a family of farmers, Rutherford excelled at university and experimented with a radio receiver about the same time as Marconi. He came to England, having won a scholarship, to pursue doctoral research on radio waves under Thomson's guidance. Instead, excited about Becquerel's discovery—and perhaps also influenced by the prominent physicist Lord Kelvin's declaration that "radio has no future"— Thomson encouraged Rutherford to investigate this new radiation.

Rutherford undertook a systematic study of Becquerel's "uranium rays" in a series of clever experiments in which he wrapped the uranium in aluminum sheets, gradually increasing the number of sheets. He realized that there must be at least two distinct types of radiation. One type, which he named alpha, could not pass through even a thin aluminum foil. The other, which he called beta, was able to penetrate a layer one hundred times thicker than the single aluminum foil. When the French physicist Paul Villard identified a third form of radiation while experimenting with radium salts in 1900, it was naturally dubbed gamma, for the third letter of the Greek alphabet.

Meanwhile, in Paris, Marie and Pierre Curie also turned their attention to Becquerel's rays. Growing up in the French capital, Pierre received his early education at home, mostly from his father, who was a doctor. After completing a physics degree at the Sorbonne, he worked as a physics instructor there, and did experiments with his brother that showed squeezing crystals could generate electricity. Later he studied magnetism for his PhD thesis, and showed that properties of magnetic materials change with temperature. Marie was born Maria Skłodowska to a family of teachers in Poland, and was determined to pursue higher education. As a teenager, she enrolled in a clandestine night school run by Polish patriots in defiance of Russian authorities. Later she worked as a governess for wealthy families in Warsaw to help her sister pay for medical school in Paris. She was able to move there herself at the age of twenty-four and enrolled at the Sorbonne. One of Marie's professors introduced her to Pierre, and the two not only became close collaborators but also developed feelings for each other. Once Pierre wrote Marie: "It would, nevertheless, be a beautiful thing in which I hardly dare believe, to pass through life together hypnotized in our dreams: your dream for your country; our dream for humanity; our dream for science. Of all these dreams, I believe the last, alone, is legitimate." At first Marie declined his proposal of marriage, but she agreed later, and the two got married in 1895. Theirs was a tight bond. As Marie wrote, "My husband and I were so closely united by our affection and our common work that we passed nearly all of our time together."

The Curies coined the term "radioactivity" to describe the phenomenon that Becquerel had discovered, and looked for other substances that showed signs of it as well. Their laboratory studies revealed that the mineral pitchblende contained two hitherto unknown elements that were even more radioactive than uranium. They named one element polonium, after Marie Curie's native Poland, and the other radium. Radium released so much heat that it felt warm to the touch. Their findings showed that radioactivity was by no means limited to uranium, and confirmed that some elements in nature can release energy spontaneously, without any external stimulation.

Sadly, they had no idea how harmful prolonged exposure to these emissions could be to the human body. Marie Curie worked for long hours in a small shed and carried test tubes containing radioactive material in her pockets. "One of our joys was to go into our workroom at night; we then perceived on all sides the feebly luminous silhouettes of the bottles or capsules containing our products. It was really a lovely sight and one always new to us. The glowing tubes looked like faint, fairy lights," she wrote, oblivious to the lethal effects that likely led to her death at age sixty-seven.

Building upon Pierre and Marie Curie's findings, Rutherford and his colleagues determined that alpha rays are made of positively charged and relatively massive particles. They later realized that alpha rays are in fact pieces of large atomic nuclei, tightly bound clumps of two protons and two neutrons. In other words, alpha particles are the same as the nuclei of helium, the second-lightest element in the periodic table, after hydrogen. These physicists confirmed that when the unstable nucleus of a heavy element such as uranium releases an alpha particle, it morphs into a slightly lighter element. Experiments done by Marie and Pierre Curie revealed that beta rays, on the other hand, are negatively charged. Becquerel and the German physicist Walter Kaufmann showed that they consist of electrons. It took several more years before physicists realized that gamma rays, the most penetrating of the radioactive emissions, are a type of electromagnetic radiation akin to X-rays and much more energetic than visible light.

Physicists now recognize that radioactivity is a fine example of that most famous of all equations, adorning countless T-shirts and coffee mugs—$E = mc^2$—in action. Derived by Einstein, this equation says that mass (m) can be converted to energy (E) and vice versa, with the speed of light (c) regulating the exchange rate. During radioactivity, when the nucleus of an atom transforms into a new configuration, some fraction of the energy locked within it is released as gamma rays or as alpha or beta particles. Energy conservation means that the total mass-energy of the end products must be identical to that of the initial grouping.

In the case of alpha and gamma decays, physicists didn't find it difficult to balance the energy accounts. That was not the case with beta decay, however. During beta decay, a nucleus appeared to release a single particle, an electron. As the British physicist James Chadwick found in 1914, the trouble was that the energy of the electron was not always the same. Sometimes the electron emerged with very little energy, while at other times it came with a whole lot. In fact, Chadwick's laboratory measurements, confirmed by others, showed that electrons emerged with a continuous range of energies up to a certain maximum value. How could that be? If energy conservation is obeyed, the electron emitted in beta decay should have the same energy every time. Some scientists wondered whether the maximum value represented the true energy released in beta decay, but a varying fraction of it somehow vanished through an unknown process.

The problem with beta decay was so persistent and so severe that Niels Bohr, the elder statesman of quantum physics, considered abandoning the hallowed principle of energy conservation altogether. As an awkward way out of the crisis, he proposed in 1930 that the conservation law might not apply strictly in the subatomic realm, but only in some average statistical sense. As he explained at a lecture in London, "We may say that we have no argument, either empirical or theoretical, for upholding the energy principle in the case of beta-ray disintegrations, and are even led to complications and difficulties in trying to do so." Even as Bohr acknowledged that "a radical departure from this principle would imply strange consequences," he argued that "in atomic theory, notwithstanding all the recent progress, we must still be prepared for new surprises."

Pauli and many others were skeptical of Bohr's suggestion, as they were not prepared to give up on energy conservation. "Do you intend to mistreat the poor energy law further?" Pauli prodded Bohr in a letter. True to form, he didn't stop there, teasing Bohr on another occasion, "What if someone owed you a great deal of money and offered to pay it back in installments, but each time the agreed-upon installment was not met? Would you consider this to be a statistical error or that something was missing?"

Indeed, even as he grappled with personal setbacks, Pauli was thinking long and hard about the predicament that physicists faced with beta decay. Finally he came to a bold resolution. Perhaps something was indeed unaccounted for: perhaps it all had to do with a ghostly particle that fled the scene of beta decay carrying the "missing" energy with it. To balance the books on energy and electric charge during beta decay, he worked out that such a particle would have to be lighter than an electron and neutral. Excited about his insight, Pauli wanted to share it with a group of top European physicists due to meet in the German city of Tübingen in early December of 1930. He was not willing to miss a winter ball in Zurich by attending the meeting, however, so he decided to send a letter to his colleagues instead.

Barely eight days after his divorce was finalized, Pauli composed a now-famous letter and addressed it: "Dear Radioactive Ladies and Gentlemen."* In it, Pauli declared that he had "hit

---

* A typed copy of the letter survives thanks to Lise Meitner, a key player in the study of beta decay, thus an important participant at the Tübingen meeting.

upon a desperate remedy" to salvage the law of energy conservation in beta decay. He proposed that electrically neutral particles might exist in the nucleus, and referred to them as "neutrons." Pauli elaborated: "The continuous beta spectrum would then be understandable on the assumption that in beta decay, along with the electron, a neutron is emitted as well, in such a way that the sum of energies of the neutron and the electron is constant."

Pauli's suggestion was daring by any measure, and he was aware of the audacity of his move. At the time, physicists knew of only three elementary particles: the proton, the electron, and the photon. Conjuring up an entirely new, as-yet-undetected particle to resolve a long-standing puzzle could seem like a cop-out to his colleagues. As he wrote to them: "I do not in the meantime trust myself to publish anything about this idea . . . I admit that my remedy may perhaps appear unlikely from the start, since one probably would long ago have seen the neutrons if they existed. But nothing ventured, nothing gained, and the seriousness of the situation with regard to the continuous beta spectrum is illuminated by a remark of my honored predecessor Mr. Debye, who told me recently in Brussels: 'Oh, it's best not to think about it at all, like the new taxes.' One ought therefore to discuss seriously every avenue of rescue. So, dear radioactive folk, put it to the test and judge."

In closing his letter, Pauli revealed his reason for missing the gathering: "Unfortunately I cannot appear in Tübingen personally, since I am indispensable here in Zurich because of a ball on the night of 6 –7 December." The dance, held at the magnificent Baur au Lac hotel overlooking the lake and the Alps, was no doubt a highlight of Zurich's social calendar that winter. Pauli's fervent desire to attend it may be a sign of his determination to get over his recent divorce as quickly as possible. Surprisingly, the usually cocky Pauli harbored some doubts about what he had sowed in the field of subatomic physics. "I have done a terrible thing. I have postulated a particle that cannot be detected. That is something no theorist should ever do," he confessed to the German astronomer Walter Baade.

The following summer Pauli traveled to several cities across the United States to deliver lectures, and at a conference in Pasadena, California, he spoke publicly about his proposed particle for the first time. "The matter still seemed to me to be quite uncertain, however, and I did not have my talk printed," he recalled later. But the news spread quickly, thanks in large part to an article that appeared in *The New York Times* on June 17, 1931, reporting that "A new inhabitant of the heart of the atom was introduced to the world of physics today when Dr. W. Pauli Jr. of the Institute of Technology of Zurich, Switzerland, postulated the existence of particles or entities which he christened 'neutrons.'"

During his lecture tour, Pauli continued to think and talk more about the hypothetical particle, while also worrying about his troubles. Despite the prohibition banning the sale of alcohol in the United States, Pauli managed to procure smuggled spirits during his travels, especially in Ann Arbor, Michigan, not far from the Canadian border. Often drinking to excess, at one house party he fell down a flight of stairs. "I recently fell (in a slightly tipsy state) so unfavorably over a stair that I have broken my shoulder and now must lie in bed until my bones are whole again—very tedious," he grumbled. A photograph shows him lecturing with his injured arm held up by a metal rod. Pauli enjoyed the trip, on the whole, even though he complained about America's puritanical tendencies and the "wretched" food at

Caltech's faculty club. He was depressed about the state of affairs in his personal life. "With women and me things don't work out at all, and probably never will succeed again," he wrote a friend from a hotel room in New York. "This, I am afraid, I have to live with, but it is not always easy. I am somewhat afraid that in getting older I will feel increasingly lonely. The eternal soliloquy is so tiresome."

From America, Pauli traveled to Italy to attend a conference on nuclear physics. It was there that he met Enrico Fermi, then a charismatic young professor at the University of Rome. Born in 1901 as the youngest of three children to a railway inspector and a schoolteacher, Fermi had shown a gift for mathematics as a child. He was especially close to his brother Giulio, and the two often built electric motors and other toys. So it was a severe blow for him when Giulio died unexpectedly during minor throat surgery. His devastated mother turned deeply melancholic, while Enrico, barely fourteen years old, became even more introverted, and immersed himself in physics textbooks, which he bought at a weekly outdoor market. One of his father's colleagues recognized young Fermi's scientific interests and encouraged him to pursue them at the University of Pisa. Just four years after enrolling at university, Fermi completed undergraduate studies as well as a doctoral thesis.

As Laura Fermi describes in *Atoms in the Family*, his thesis defense was a bittersweet occasion: "The eleven examiners in black togas and square-topped hats were sitting in solemn dignity behind a long table. Fermi, also in a black toga, stood in front of them, and he started talking with cool, deliberate assurance. As he proceeded, some of the examiners repressed yawns, some sent their brows up in wonder, others relaxed and did not try to follow. Clearly, Fermi's erudition was above their comprehension. Fermi received his degree *magna cum laude.* But none of the examiners shook hands with him or congratulated him . . ." Afterward, thanks to a government scholarship, Fermi was able to travel to Germany and the Netherlands to work with prominent physicists of the time. Offered a professorship in Rome at age twenty-six, he quickly gathered a group of talented students around him, many of whom went on to make their own mark in physics, including Bruno Pontecorvo and Ettore Majorana, whom we'll hear more about. Fermi's colleagues referred to him as "the Pope" because he was a natural leader and seemed infallible.

Fermi was intrigued by Pauli's solution to the puzzle of beta decay. As Pauli recalled, he "immediately expressed a lively interest in my idea and a very positive attitude toward my new neutral particles." Both found Bohr's alternative proposal to give up on energy conservation unacceptable because that would violate a basic law of physics that they cherished. The following year, James Chadwick discovered a previously unknown neutral particle in the atomic nucleus. Nearly the same mass as the proton, it was much too massive to match Pauli's prediction. Since Chadwick named his particle the neutron, the same term that Pauli had used previously to describe his hypothetical particle, Fermi introduced a new name for the latter. He called it the neutrino—Italian for "little neutral one"—and physicists began to use this new name.

Back in Zurich, Pauli's drinking, smoking, and womanizing reached new extremes. He spent a lot of time at bars, got into fights, and quarreled with colleagues. He exhibited sudden mood swings, and was on the brink of a nervous breakdown toward the end of 1931. Heeding his father's advice, Pauli consulted the eminent psychoanalyst Carl Jung. He read Jung's works, attended

his talks, and made an appointment to see the famous therapist. Jung wrote about his initial encounter with Pauli: "When
the hard-boiled rationalist [Pauli] . . . came to consult me for
the first time, he was in such a state of panic that not only he
but I myself felt the wind blowing over me from the lunatic
asylum!" To Pauli's surprise, given his difficulties with women,
Jung asked him to contact one of Jung's young female pupils,
Erna Rosenbaum, for therapy. Pauli complied, wanting "nothing to be left untried." Over the next several months, he recounted hundreds of his dreams to Rosenbaum during therapy
sessions and in letters, even after she left Zurich for Berlin.
Later Jung himself took over Pauli's case. Pauli worked with
Jung for two intense years, describing his often-elaborate dreams
to the therapist, who in turn provided sophisticated analyses of
their motifs and symbols.

In 1933, Pauli met Franca Bertram, a cultured German
woman who had traveled widely and worked in Zurich as the
manager of a Russian orchestra. They got married the following
year, and she became his companion for the rest of his life.
Franca distrusted psychoanalysts, and soon after the wedding,
Pauli ended his consultations with Jung. But the two men
continued to correspond for decades, exploring their common
interests not only in psychology but also in mysticism and nu-
merology. Jung drew upon Pauli's rich tapestry of dreams for
his lectures and writing, taking care not to reveal the identity of
his client-turned-friend. Many of Pauli's physicist friends seem
not to have known about this remarkable friendship while he
was alive. Given the emotional turbulence of this period, it is
no wonder that many years later Pauli described the neutrino
as "that foolish child of the crisis of my life."

Meanwhile, back in Rome, after his meeting with Pauli,

Fermi continued to ponder the mystery of beta decay. At a major conference on atomic nuclei he attended in Brussels in the fall of 1933, it was again a key topic of discussion, and in the months following the meeting, Fermi was able to formulate a clear mathematical description of beta decay within the framework of quantum mechanics. His theory assumed that the nucleus consisted of heavy particles, namely, protons and neutrons, as the quantum pioneer Werner Heisenberg had suggested. Fermi's theory postulated that during beta decay a neutron morphs into a proton, which remains in the nucleus, while an electron is released along with a neutrino, as Pauli had proposed. Fermi made it clear that the neutrino did not reside in the nucleus to begin with, but was created during beta decay. He compared the results of his theoretical calculations with experimental data, and concluded that "the rest mass of the neutrino is either zero or in any case very small with respect to the mass of the electron."

Furthermore, Fermi's theory foreshadowed a new fundamental force of nature, what we now call the weak force, which operates in the subatomic realm. Two of the four known forces, gravity and electromagnetism, act over large distances, so we are familiar with them in our everyday experience. For example, we experience the Earth's gravity when we lift something heavy, and we sense the pull of a magnet on a refrigerator door. The other two forces, strong force and weak force, however, act over minuscule distances inside atoms. The strong force is what binds protons and neutrons together in the atomic nucleus. The weak force governs the processes of radioactivity, including beta decay.

Fermi submitted his paper on the theory of beta decay to *Nature* in 1934. The journal's editors were not impressed, however: they rejected his pièce de résistance, claiming that "it contained speculations too remote from reality to be of interest to the reader." So Fermi pressed on, and submitted his paper to two other journals instead: one based in Italy, because Mussolini's Fascist government required publication in Italian, and the other based in Germany so that scientists abroad could read it. Both journals published it. The paper confirmed Fermi's reputation as a sharp theorist who could see through the clutter, and it remains a classic in the history of physics. As Christine Sutton put it so eloquently in her book *Spaceship Neutrino*, "If Pauli's letter to the 'radioactive ladies and gentlemen' marked the conception of the neutrino, then Fermi's paper on beta decay surely heralded its birth." The problem, however, was that nobody knew how to take a snapshot of the elusive newborn.

Next, Fermi turned his attention to experiments on "artificial radioactivity" to better understand the phenomenon of nuclear transformation. By then, Marie Curie's daughter Irène Joliot-Curie and her husband Frédéric Joliot had shown that bombarding certain nuclei with alpha particles produces new radioactive species that subsequently decay. Though turning lead into gold was not within their grasp (nor was it their objective), in some sense these scientists had realized the fanciful dreams of ancient alchemists who wanted to turn common metals into rare, valuable varieties.

While the Joliot-Curies had used alpha particles as the projectiles, Fermi tried neutrons instead. He found that bombarding atoms with slow-moving neutrons was especially effective in creating radioactive products. Legend has it that his insight was based on the fact that the experiment recorded much stronger radioactivity when conducted on a wooden tabletop than on a marble tabletop. His colleagues found the difference puzzling, but Fermi figured out that it was due to light atoms in the wood slowing the neutrons down. Building on Fermi's findings, other scientists experimented with hitting uranium with slow neutrons. In 1938 physicists in Germany succeeded in splitting the uranium nucleus roughly in half, opening up the prospect of unleashing the enormous energy trapped in atomic nuclei.

That same year, Fermi was given a Nobel Prize for his work on slow neutrons. His trip to Stockholm to accept the award gave him a chance to leave Italy with his Jewish wife and two children, as anti-Semitism was on the rise in his home country, and to move with them to the United States. After a few years at Columbia University in New York, Fermi moved to the University of Chicago. It was there in 1942, in a squash court under the university's stadium, that a group of physicists under Fermi's direction achieved the first controlled nuclear chain reaction, the essential first step to harnessing nuclear energy. One of his colleagues reported the success to James Conant, chair of the National Defense Research Committee in Washington, in a coded phone call, saying that "the Italian navigator has landed in the New World." Conant asked, "How were the natives?" The answer came, "Very friendly." Fermi's work in Chicago set the stage for atomic bombs, nuclear power, and the ultimate experimental discovery of neutrinos.

Meanwhile, back in Europe, Fermi's splendid paper on beta decay had not been forgotten, even though many physicists had moved on to work on other topics. In fact, the paper had prompted some theorists and experimentalists to keep thinking about ways to pin down the ghostly neutrinos. Two German physicists, Hans Bethe and Rudolf Peierls, considered an interesting possibility. Since a neutrino is released in beta decay, could one be absorbed in a reverse process, the same way that photons are emitted and absorbed by atoms? They found that the odds of a neutrino being absorbed by an atom are puny. The two theorists concluded in a brief note to *Nature* that there was "no practically possible way" of observing the neutrino.

Their somber calculation did not deter plucky neutrino hunters, however. At Cambridge University, James Chadwick and D. E. Lea tried to measure the penetration levels of neutrinos by placing varying amounts of lead between a sample of radium and a detector. The researchers thought that the lead would slow down the neutrinos, making these elusive particles easier to detect. However, their findings showed that on average a neutrino could travel more than ninety *miles* through the air before interacting with a single atom. The British physicist Maurice Nahmias performed an even more sensitive experiment the following year, hoping to trap neutrinos. He set up his apparatus at the Holborn subway station in London, about a hundred feet below the surface, to reduce unwanted background radiation from energetic particles that arrive from space. He too failed to detect the evasive neutrino. His measurements implied that neutrinos released in beta decay could pass unhindered through the entire Earth. What's more, Nahmias's subway experiment was a harbinger of things to come: deploying detectors underground, to minimize confusion caused by unrelated sources of noise, is now a common practice in particle physics.

Neutrinos, it seemed, were destined to be poltergeists, mysterious phantoms that stole energy in beta decay but otherwise remained intractable. Since they are electrically neutral, unlike protons and electrons, they cannot be traced through electromagnetic means. They do not feel the strong force either, and the chance of a neutrino interacting with an atomic nucleus via the weak force is extremely small.

As a result of discouraging theoretical estimates and the failure of experimental attempts, a growing chorus of skeptics doubted that the neutrino would ever be detected. Among them was the Nobel laureate Paul Dirac, who predicted the existence of antimatter, as we will learn in chapter 7. He was favorably inclined toward neutrinos at first, and praised Fermi's theory of beta decay in a letter to a colleague in 1934, in which he noted that "neutrinos seem to provide the only escape from non-conservation of energy and until something else turns up one should not be unsympathetic to them." Barely two years later, he seemed to have had a change of heart, dismissing the neutrino as an "unobservable particle."

By the end of the 1930s, other scientists shared Dirac's doubts about Pauli's neutrino hypothesis. The British astrophysicist, expositor of relativity theory, and science popularizer Arthur Eddington reflected the skepticism of the time when he noted in his book *The Philosophy of Physical Science*: "Just now nuclear physicists are writing a great deal about hypothetical particles called neutrinos supposed to account for certain peculiar facts observed in beta-ray disintegration . . . I am not much impressed by the neutrino theory. In an ordinary way I might say that I do not believe in neutrinos . . . Dare I say that experimental physicists will not have sufficient ingenuity to make neutrinos?" He didn't rule out the possibility that neutrinos might exist, however, adding, "Whatever I may think, I am not going to be lured into a wager against the skill of experimenters under the impression that it is a wager against the truth of a theory. If they succeed in making neutrinos, perhaps even in developing industrial applications of them, I suppose I shall have to believe—though I may feel that they have not been playing quite fair."

Ironically, it was Pauli who wagered a case of champagne that no one would be able to detect the neutrino experimentally. Perhaps he was having second thoughts about the poltergeist he had conjured up. Or perhaps he thought that betting against his own brainchild was a good way to fend off his critics. Either way, Pauli's bet remained unclaimed for a quarter of a century, as physicists in Europe and North America turned their attention to the Second World War. Their wartime efforts to exploit the energy trapped within atomic nuclei not only led to the deadliest weapons ever known to humankind but also resulted in tremendous new sources of neutrinos.

THREE

GHOST CHASING

The next chapter of the neutrino saga was connected intimately to the harnessing of nuclear energy during the Second World War and its aftermath. As early as 1938, two German chemists, Otto Hahn and Fritz Strassmann, found that bombarding uranium with neutrons produced the much lighter element barium. Their former colleague Lise Meitner and her nephew Otto Frisch, who had both moved to Scandinavian countries as Nazis targeted Jewish academics in Germany, correctly recognized that the transformation was the result of splitting the heavy uranium nuclei and dubbed the process "fission," borrowing a word that biologists use to describe the subdivision of living cells. In a paper published in *Nature*, they wrote: "It seems therefore possible that the uranium nucleus has only small stability of form, and may, after neutron capture, divide itself into two nuclei of roughly equal size." Even before the paper appeared in print, Niels Bohr brought news of the discovery to America, which prompted Enrico Fermi and others to carry out fission experiments themselves, and also raised the specter of atomic bombs.

Soon a number of prominent scientists began warning the

Allied nations of the potential dangers if Nazi Germany were to acquire nuclear weapons. They acted out of the fear that whichever side unleashed the power of atomic nuclei first would gain an overwhelming dominance over the other in the conduct of the war. At the urging of several colleagues, Albert Einstein signed a now-famous letter to President Franklin D. Roosevelt, in which he declared that "it may become possible to set up a nuclear chain reaction in a large mass of uranium, by which vast amounts of power and large quantities of new radium-like elements would be generated. Now it appears almost certain that this could be achieved in the immediate future. This new phenomenon would also lead to the construction of bombs . . ." He encouraged the president to secure a supply of uranium ore, hinting that Germany had already taken steps to do so, and to bolster contacts with nuclear physicists to keep abreast of the latest research findings.

Three years later, the United States government, with the support of Canada and the United Kingdom, initiated the top-secret Manhattan Project and recruited many of the nation's top theoretical and experimental physicists to work on it, with the goal of harnessing the power of chain reactions to develop atomic bombs. However, Wolfgang Pauli, who was now working with Einstein at the Institute for Advanced Study in Princeton, was not invited to join because of his German passport (Germany had annexed Austria by then and he had failed to secure Swiss citizenship). While scientists at the newly created Los Alamos Laboratory in New Mexico led the weapon design, assembly, and testing, others at various sites around the country were charged with enriching uranium and producing plutonium for use in the bombs. At first the project was a modest undertaking, but eventually it employed well over 100,000 people. The Los

Alamos scientists detonated the first nuclear device, code-named "the gadget" and mounted on a 100-foot tower, in the New Mexico desert on July 16, 1945. As they watched the glowing mushroom-shaped cloud rise into the sky and heard the roar of the shock wave, the witnesses of the test realized the awesome destructive power of the weapon they had built—and some came to regret their role in its development. Just a few weeks later, the American forces dropped two atomic bombs on the Japanese cities of Hiroshima and Nagasaki, causing death and devastation on a horrific scale and precipitating the war's end. In the years that followed, the United States and other countries tested many more bombs as they sought to develop ever more potent military capabilities.

Aside from their destructive potential, atomic bombs served as spectacular sources of neutrinos. Nuclear fission during the explosions produced unstable nuclei that decayed subsequently, releasing staggering bursts of these particles. The enormous numbers improved the prospects of detecting neutrinos experimentally, because the larger the flux of particles, the higher the chance that some would register in a detector. But neutrinos were not foremost in the minds of the world's top physicists on both sides of the Atlantic as many of them focused on the war effort, including Enrico Fermi, who was very much involved in the Manhattan Project. One of Fermi's Italian protégés, Bruno Pontecorvo, however, was not directly involved in weapons research, and he developed the greatest insights into the nature of the elusive neutrinos and how to trap them, while spinning a web of intrigue by his own evanescence.

Born to well-to-do Jewish parents near Pisa, Pontecorvo grew up not far from the city square where Galileo had conducted his legendary experiments with falling bodies in the

Bruno Pontecorvo
(AIP Emilio Segrè Visual Archives,
Physics Today Collection)

seventeenth century. Pontecorvo's prosperous family owned a textile factory with a large number of workers, took splendid vacations in the summer, and employed private tutors to educate their children. As an adolescent, Pontecorvo excelled in tennis and science at school. After high school, he enrolled at the University of Pisa to pursue engineering, but moved to the University of Rome two years later to study physics under Fermi, who was already gaining an international reputation as a scientific leader. Italy was in turmoil at this time, with the rise of the Fascists, but young Pontecorvo wasn't politically active, even though some of his family members were pacifists and leftists. Mussolini's government strongly supported scientific research, considering it essential for industrial progress of the country. They encouraged large companies to invest in innovation, established the Italian national research council, and increased state funding for laboratories. Fermi benefited from the patronage and took care to keep politics out of his laboratory so as not to offend government officials, even though he later distanced himself from the regime.

After graduating with high marks, Pontecorvo joined Fermi's research group. He was clumsy around laboratory equipment at first, but mastered experimental techniques quickly, and in 1934 he played a key role in the discovery that bombarding atoms with slow-moving neutrons enhances reaction rates. Pontecorvo and his colleagues shared credit for the finding as coauthors of a scientific paper and co-owners of a patent for its possible commercial application. Two years later, he moved to Paris to work with Pierre and Marie Curie's daughter Irène Joliot-Curie and her husband Frédéric Joliot, who were experimenting with nuclear transformations. Paris was a hotbed of left-wing activism at the time and a favorite destination for political refugees fleeing from totalitarian regimes elsewhere in Europe. So it's no surprise that Pontecorvo was exposed to socialist ideas in Paris and made a number of friends active in leftist politics, particularly through his brother Gillo, who had followed him to Paris. Gillo was a dedicated Marxist and a member of the Italian Communist Party, and he went on to direct the iconic anticolonial film *The Battle of Algiers* decades later. Discussions with his brother and friends may have led Pontecorvo to consider his social and political responsibilities as a scientist. Also while living in Paris, Pontecorvo met Marianne Nordblom, a Swedish student of French literature, at a hostel, and the two decided to move in together. Soon she gave birth to their first son.

With anti-Semitism on the rise in his home country and the Nazis advancing into France, Pontecorvo and Nordblom got married and sought refuge in the United States in 1940, just as Fermi had done two years earlier. He found a job with an oil company in Oklahoma, where he used his expertise in nuclear physics to develop new technologies for oil prospecting. Although he was now based in the United States, and despite his close connection to Fermi, Pontecorvo was not invited to join the Manhattan Project, perhaps because the security services were wary of his socialist leanings. Instead, he joined a British-Canadian effort to build a nuclear reactor near Chalk River, Ontario, moving with his family to Canada in 1943.

It was while in Canada that Pontecorvo turned his attention to neutrinos. He was, of course, familiar with his mentor Fermi's theory of beta decay, which predicted the release of a neutrino along with an electron as a neutron morphed into a proton. Contrary to prevailing opinion, however, Pontecorvo firmly believed that physicists should be able to detect neutrinos with the right experimental setup. The odds of a particular neutrino interacting with a detector were extremely small, but Pontecorvo thought that if there were many trillions of particles reaching a detector every second, it should be possible to capture a few. The first step toward this goal, he noted, was to identify a copious source of neutrinos. He knew that even a very large chunk of radium would not release enough neutrinos through beta decay to do the trick. But a nuclear reactor, he reasoned, should produce trillions of these particles each second. As he recalled almost forty years later, "It occurred to me [in 1946] that the appearance of powerful nuclear reactors made free neutrino detecting a perfectly decent occupation." Given his intimate knowledge of nuclear power generation, this insight is not too surprising. But he went further: he outlined how to go about *trapping* these ghostly particles.

Pontecorvo knew that, according to Fermi's theory, two things should happen when a neutrino hits an atomic nucleus: one, the neutrino picks up a negative charge and turns into an electron, and two, the nucleus gains a positive charge to balance out the books. In other words, the atom hit by the neutrino should turn into an atom of a different element, one that is next on the periodic table, by transforming one of the neutrons in its nucleus into a proton. If this new atom was radioactive, Pontecorvo realized, its presence would be revealed when it decays and emits radiation. So he worked out the practical requirements for setting up such an experiment. First, he had to identify a target material that was relatively cheap and easy to obtain, because lots of it would be needed to make a sufficiently large detector. Second, the target material had to be one that should turn into a radioactive substance upon absorbing neutrinos. Third, the radioactive product should not decay too quickly, before there was time to measure it. Based on these considerations, Pontecorvo proposed using a huge tank of dry-cleaning fluid, or carbon tetrachloride, which contains atoms of chlorine. He knew that according to Fermi's theory, when a chlorine atom collided with a neutrino, it should turn into argon, a radioactive element that is chemically inert. The subsequent decay of the argon atom would signal that a neutrino had struck its predecessor. He had come up with a clever way to prove the existence of the otherwise undetectable poltergeist.

Yet Pontecorvo didn't have the chance to hunt for neutrinos himself. He applied for British citizenship, and moved to England with his wife and three sons in 1948. It was his fourth move in twelve years. Within two years of starting work at an atomic research laboratory in Harwell near Oxford, Pontecorvo had some immediate concerns on his mind that eclipsed his desire to crack the case of the elusive neutrinos. Suspicions about his Communist leanings attracted growing attention from the FBI in the United States and MI5 in Britain. There was pressure for Pontecorvo to quit his job at Harwell, where he may have had access to sensitive military research, and take up an academic position instead.

As Pontecorvo was well aware, tensions stemming from the Cold War were on the rise. The Soviet Union had carried out a successful atomic bomb test in 1949, surprising many in the West by its technological prowess and challenging American nuclear supremacy. In response, President Harry Truman decided to develop even more powerful hydrogen bombs, which involve turning hydrogen isotopes into helium through nuclear fusion (but do still need fission reactions to trigger the explosion). Fears of Communist influence and spies raged in Washington, of course, with Senator Joseph McCarthy at the helm of a notorious witch hunt that targeted anyone suspected of disloyalty. In Britain, Klaus Fuchs, a physicist at the same Harwell laboratory where Pontecorvo worked, confessed to spying for the Russians while employed by the Manhattan Project, and was swiftly convicted in a high-profile trial. In the United States, the FBI investigated others who worked at Los Alamos, including the colorful physicist Richard Feynman, who was known for his brilliance as well as his mischievous deeds, including picking locks of secure cabinets just to prove that he could. Feynman was cleared, but the federal agents found that a chemist named Harry Gold had acted as a courier for Fuchs. Gold, in turn, led them to Ethel and Julius Rosenberg, who were arrested on suspicion of espionage along with Ethel's younger brother David Greenglass, a former machinist at Los Alamos. The spy revelations and arrests contributed to the prevailing atmosphere of suspicion and paranoia.

Adding to Pontecorvo's worries, a battle was brewing between the U.S. government and Pontecorvo's former Italian colleagues over the slow-neutron patent that resulted from the work of Fermi's group in Rome. These researchers claimed a financial stake in the production of plutonium, which used a technology that relied on slowing down neutrons with graphite. In 1950, one of the patent holders filed a lawsuit seeking compensation from the U.S. government. The prospect of being drawn into a public legal dispute, just as the security checks into his background were intensifying, was probably too much for Pontecorvo.

When the news of the patent lawsuit reached Europe, Pontecorvo was on a summer holiday in Italy with his wife and sons. After the vacation, instead of returning to Britain, where Pontecorvo was to take up a professorship at Liverpool University, the family traveled from Rome to Stockholm on the first of September. Curiously, even though Pontecorvo's wife's parents lived in the city, the family didn't make contact with them upon their arrival, and the next day, they caught a flight to Helsinki. This is where the paper trail ends. There is no record of the family's whereabouts in Helsinki and no evidence that they crossed the border into another country.

A newspaper in Italy was the first to report on their sudden disappearance under mysterious circumstances, perhaps tipped off by British investigators making inquiries in Rome. From the very beginning, the security services suspected that Pontecorvo had fled with his family to the Soviet Union. Indeed, some of his friends and family members were known to be leftists, if not Communist sympathizers. The newspapers at the time played with this suspicion to great effect, and a low-key scientist whom the public had not previously heard of became a worldwide sensation practically overnight. "Atom Man Flies Away" read the banner headline of the *Daily Express* in London, while *The Manchester Guardian* declared "Atomic Expert Missing." Even the BBC surmised that Pontecorvo had defected behind the Iron Curtain, and reported that "British intelligence service MI5 has been brought into the hunt for the missing atomic scientist Bruno Pontecorvo who has not been seen for about seven weeks." News outlets around the world were keen to suggest possible motives for defection. Indeed, *The Sydney Morning Herald* quoted a physicist who knew Pontecorvo, "He is one of the most all-round men in atomic science, and certainly one of the very best in Britain. No doubt, a man like him would be highly useful to the Russians." Perhaps, as many media reports pointed out, the Pontecorvo family slipped out of Finland by train or ship from a Russian military base near Helsinki. Needless to say, the commotion surrounding Pontecorvo's disappearance added to the Cold War intrigue of the period.

Many people feared that Pontecorvo had been a Russian spy all along. Others thought he had run away because the secret services were on his tail, or because he wanted to escape from anti-Communist hysteria in the West. Still others speculated that Russian agents had somehow coerced Pontecorvo into switching allegiances. Whatever the reason, Pontecorvo's disappearance was an extraordinary event, and one that both the American and British intelligence authorities downplayed in their public statements. They were likely embarrassed by the apparent defection of a prominent physicist, especially coming on the heels of the Fuchs spy revelations, and wanted to avoid feeding a widespread frenzy. As science historian Simone Turchetti recounts in *The Pontecorvo Affair*, the British minister of supply George Strauss went so far as to pledge to Parliament that Pontecorvo had not been involved in classified military work.

What's more, the Foreign Office sought to reassure British diplomats abroad through a confidential telegram, which asserted that "Dr. Pontecorvo was engaged in work of a non-secret nature at Harwell, and, although he may be of use to the Russians in the field of basic research, it is not thought that he knows anything of value regarding atomic weapons."

Confirming many people's suspicions, Pontecorvo surfaced in Moscow five years later. He gave his own version of the reasons behind his defection—a desire to counterbalance Western domination and prevent another world war—and he offered assurances that he was only interested in working on peaceful uses of atomic energy. The Soviet Union had given him a warm welcome and offered him a job at a nuclear research institute not far from Moscow, honoring him with the Stalin Prize and other accolades.

Years later, Pontecorvo revealed that the family had been smuggled out of Helsinki by the Soviet embassy. He was hidden in the trunk of one of the two cars that transported the family across the Finnish border. We do not know for sure whether Pontecorvo contributed to the development of Soviet atomic weapons, either through espionage before his defection or through his expertise afterward, and six decades later, the FBI and MI5 files on his case remain under wraps. Simone Turchetti, for one, doesn't think he was a spy, but the historian also believes that Pontecorvo's expertise in geophysical prospecting may have helped the Russians locate uranium reserves that they needed for making bombs. Whatever Pontecorvo worked on while living behind the Iron Curtain, he retained a keen interest in neutrinos, and was the first to theorize about their chameleonlike nature, as we will find out soon.

Meanwhile, back in the United States, many of the physicists involved in the Manhattan Project had returned to universities. Among the few who remained at Los Alamos in the early 1950s was a thirty-three-year-old theorist by the name of Fred Reines. Growing up in small towns in New Jersey and New York, Reines was a talented singer and a Boy Scout as a child. His earliest memory of fascination with science involved staring through a curled hand at a window at twilight, during a moment of boredom at religious school, and being captivated by the diffraction of light. He also enjoyed building crystal radios—those simple receivers popular in the early days of radio—from scratch, and revealed in his high school yearbook that his ambition was "to be a physicist extraordinaire." While pursuing engineering at the Stevens Institute of Technology, Reines sang in a chorus and performed solos in major pieces such as Handel's *Messiah*. As he recalled, "Between college and graduate school, I even thought briefly about pursuing a professional singing career." Even before he completed his PhD thesis at New York University in 1944, he was recruited to Los Alamos, where he got involved in bomb tests during and after the war, including those conducted at Bikini and Eniwetok atolls in the Pacific. As a theorist, Reines focused on developing a better theoretical understanding of the effects of nuclear blasts, including how the blast waves propagate through the air. He continued his musical pursuits in his spare time, singing with the town choir and acting with a local drama club, and later even performed with the chorus of the Cleveland Symphony Orchestra.

In 1951, Reines asked his boss at Los Alamos for a research leave to focus on fundamental physics. As Reines later recalled, reflecting on this turning point of his life, his boss agreed and he "moved to a stark empty office, staring at a blank pad for several months searching for a meaningful question worthy of a

Frederick Reines (AIP Emilio Segrè Visual Archives)

life's work." As he thought about interesting projects to pursue, it occurred to him that a nuclear bomb should produce a prodigious burst of neutrinos, and could therefore come in handy for detecting these spooky particles. He knew that fission of atoms through a chain reaction produces a multitude of unstable nuclei, which decay in turn through the beta process and emit neutrinos. On average, each fission event results in six neutrinos, so the upshot is a tremendous burst of them that is released at once. According to Reines, "Some hand-waving and rough calculations led me to conclude that the bomb was the best source . . . I thought, well, I must check this with a real expert."

As it happened, Enrico Fermi was visiting Los Alamos in the summer of 1951, and Reines summoned up the courage to speak with him. Fermi agreed with Reines that a nuclear bomb was indeed a terrific neutrino source, but the real problem was that Reines had no idea how to make a suitable detector to trap the particles. After some thought, Fermi admitted that he didn't either. "Coming from the Master, that was very crushing," Reines remembered. His dreams of becoming a neutrino hunter were put on hold until he could figure out a way forward.

Luckily, a happy accident intervened soon enough. Later in 1951, Reines was on his way to Princeton when his plane was grounded at the Kansas City airport because of engine trouble. His Los Alamos colleague Clyde Cowan was with him. Cowan had earned a Bronze Star for his work on radar during the war, and then went to graduate school on the G.I. Bill, completing a doctorate in physics. He joined Los Alamos in 1949. As the two stranded scientists passed the time, wandering about the airport, they decided to work together on a problem in physics that they both found challenging. Reines proposed they focus on neutrinos, and Cowan agreed. "He knew as little about the neutrino as I did but he was a good experimentalist with a sense of derring do. So we shook hands and got off to working on neutrinos," according to Reines. As he put it, "So why did we want to detect the free neutrino? Because everybody said, you couldn't do it. Not very sensible, but we were attracted by the challenge." Their experience with tackling formidable military projects may have had something to do with their gutsy attitude. "Bomb testing was an exercise in thinking big, in the 'can do' spirit . . . [It] permeated our thinking. Whenever we thought about new projects, the idea was to set the most interesting (and fundamental) goal without initial concern as to feasibility or practical uses. We could count on the latest technology being available to us at Los Alamos . . . and that fact fed our confidence," Reines explained.

Reines and Cowan knew that most neutrinos pass right

through matter. But if there were a sufficiently large number of these particles, a few would interact with atomic nuclei by chance. With that in mind, the two of them chose to focus on a particular reaction in their quest: when a proton absorbs a neutrino, Fermi's theory predicts, it turns into a neutron while emitting an anti-electron, also called a positron.* They knew that if they detected this positron, they could establish that a neutrino was present at the scene. What they needed was a way to register the positrons. Luckily for them, researchers had recently identified organic liquids that scintillate (or produce tiny flashes of light) when a charged particle passes through them. So Reines and Cowan planned to fill a large vat with scintillator fluid and mount several photomultiplier tubes on the inside of its walls to record the flashes resulting from the positrons. They intended to suspend this detector from the top of a vertical shaft dug into the ground, just 130 feet from a tower on which they would explode a 20-kiloton fission bomb. As Reines later admitted in his lecture at the Nobel Prize ceremony, "The idea that such a sensitive detector could be operated in the close proximity . . . of the most violent explosion produced by man was somewhat bizarre, but we had worked with bombs and felt we could design an appropriate system."

With this design, Reines and Cowan could allow the detector to fall freely in a vacuum inside the shaft for a few seconds while the shock wave from the explosion passed by. Then the vat would land on a thick layer of foam rubber and feathers, which would soften its touchdown, and register the positrons

---

* In fact, this process, called "inverse beta decay," involves the interaction between a proton and an antineutrino to produce a neutron and a positron. Reines and Cowan did not know at the time that there might be a distinction between a neutrino and an antineutrino.

emitted by the bomb's fission fragments as the fireball launched into the sky. In a later report, Cowan described their plan for retrieving the experimental data after the bomb test: "We would return to the site of the shaft in a few days (when the surface radioactivity had died away sufficiently) and dig down to the tank, recover the detector, and learn the truth about neutrinos!" As outlandish as their scheme might sound today, it was a clever setup, and the director of the Los Alamos laboratory gave it the go-ahead. As Reines explained, the approval process was fairly straightforward at the time: "Life was much simpler those days—no lengthy proposals or complex review committees."

After Reines and Cowan had arranged for workers to come to the site and dig a massive hole and had started to build the detector, their colleague Hans Bethe asked them how they expected to distinguish a genuine neutrino event from the other radiation that he knew would be emitted by the bomb. While they tried to come up with answers to his question, Reines and Cowan arrived at an even better way to do the experiment: instead of a one-off bomb explosion, they could use a controlled nuclear reactor as the neutrino source. Even though a reactor produced far fewer neutrinos per second than a bomb would, the number was still in the trillions per square inch. They believed that was more than enough to securely detect the neutrino. The reason had to do with a more reliable way to distinguish neutrinos from other unrelated "background" events, such as cosmic rays. Instead of registering only the positron, Reines and Cowan realized they could also measure the neutron that was produced as a result of a neutrino interacting with matter. They knew that the positron would hit an electron in the scintillator fluid, annihilating both almost instantly and producing gamma rays that the photomultiplier tubes could record. The neutron, on the other hand, would zigzag through the fluid, like a drunkard staggering through a crowd, losing energy as it collided with one nucleus and then another, until it was eventually absorbed. The nucleus that captured the neutron would emit the excess energy as gamma rays. As Reines and Cowan were aware, there is a characteristic time for the neutron's random walk: 5 microseconds. That means there is a precise delay between the two gamma ray bursts, the first one from the positron's annihilation and the second from the neutron's absorption. If their experiment recorded two flashes coming precisely 5 microseconds apart, it would constitute the unmistakable signal of a neutrino. Reines and Cowan could then conclude that flashes due to other particles hitting the detector at random times were simply background noise.

The new plan that Reines and Cowan devised was much more practical and a lot safer than relying on a large explosion from a bomb, of course. They were either unaware of Pontecorvo's 1946 report, in which he had proposed using a reactor as a source of neutrinos for detection, or they had forgotten about this sensible suggestion. The two physicists shared their revised strategy to track the "slippery particle" in a letter to Fermi; after all, they had no fear of being scooped by someone else, since "neutrino detection was not a popular activity in 1952," as Reines later put it. Fermi endorsed their revised course of action: "Certainly your new method should be much simpler to carry out and have the great advantage that the measurement can be repeated any number of times."

Encouraged by Fermi's support, Reines and Cowan moved ahead, and in 1953, they constructed a cylindrical tank holding 300 liters (80 gallons) of scintillator liquid, viewed by ninety photomultiplier tubes dotting the inside walls. Thick layers of paraffin, borax, and lead surrounded the tank, shielding it from stray neutrons and gamma rays coming from the reactor. They called the experiment "Project Poltergeist," which was a fitting name for the particle they were chasing. The detector was set up near a nuclear reactor in Hanford, Washington, which had been built during the war to produce plutonium for atomic bombs. Decades later, Reines reminisced about the excitement, but also the fatigue, associated with their chase: "Those days at Hanford were both stimulating and exhausting. For a few months we stacked and restacked several hundred tons of lead and boron-paraffin shielding. We worked around the clock as we struggled with dirty scintillator pipes . . ."

They recorded the first hints of a signal within months. The detections were not as clear-cut as Reines and Cowan had hoped, however: their detector registered events even when the reactor was turned *off*. It turned out that cosmic ray particles coming from outer space were responsible for mimicking the characteristic double flashes that they expected to see from neutrinos interacting with matter. Still, Reines and Cowan saw that the count rate was higher when the reactor was on, so they believed they had at least provisionally detected neutrinos. In a brief note published in the *Physical Review*, they were cautious about these results, stating, "It appears probable that this aim [of detecting neutrinos] has been accomplished although further confirmatory work is in progress." Cowan later described the situation best: "We felt we had the neutrino by the coattails, but our evidence would not stand up in court."

Despite their well-warranted caution, the news of their results made it to the popular press. "Atom Team Sees 'Ghost' Particle," *The New York Times* reported. *Time* and *Scientific*

American also published articles about the results, while a press release from Science Service made bold claims about how textbooks would have to be revised, declaring that "[The] poltergeist of modern physics . . . has been caught." When the news reached Wolfgang Pauli in Zurich, he and a group of friends promptly hiked up a nearby mountain with a panoramic view of the city for a celebratory dinner. Apparently, a jubilant Pauli was a little wobbly on the trip down the mountain later that night.

But being the hard-nosed scientists they were, Reines and Cowan were not satisfied with their result, which was indeed tentative at best. They decided to try again with a more sensitive experiment at the newly completed reactor at the Savannah River Site in South Carolina, which was much more powerful than the one they had used at Hanford. With the help of several colleagues, they redesigned the experiment from top to bottom, to distinguish *true* neutrino events from spurious signals caused by cosmic ray particles, and they equipped the experiment with multiple scintillator tanks. The new apparatus, completed in late 1955, weighed in at some ten tons. Located in a basement below the nuclear reactor, it was shielded from neutrons produced by the reactor and from cosmic rays. Over five months, the scientists recorded hundreds of hours of data with the reactor turned on and off for comparison. The apparatus registered five times as many pairs of flashes separated by a few microseconds when the reactor was on than when it was off. By the summer of 1956, after extensive tests and checks, the team was convinced that they had definitively detected neutrinos.

As Reines recalled, "It was a glorious feeling to have participated so intimately in learning a new thing, and in June of 1956 we thought it was time to tell the man who had started it all when, as a young fellow, he wrote his famous letter in which he postulated the neutrino." Reines was of course referring to Pauli, and he and Cowan telegraphed Pauli the following news: "We are happy to inform you that we have definitely detected neutrinos from fission fragments . . ." The message reached Pauli at a conference in Geneva. He interrupted the proceedings and read the telegram out loud.

Pauli wrote back to Reines and Cowan the next day, and, true to his character, he quipped: "Everything comes to him who knows how to wait." But Pauli's message failed to reach them. A copy exists in the Pauli archives with a note from his secretary that it was indeed sent by "night letter" (a telegram sent overnight at reduced rates for delivery the next day). Pauli's response would have been a prescient statement, since it took four decades for the Nobel committee to recognize the discovery of the neutrino. Reines was awarded a share of the 1995 physics prize. Cowan had died twenty-one years earlier, so he missed out on the honor. Years after the discovery, Reines confronted the theorist Hans Bethe, who had asserted in his 1934 paper with Rudolf Peierls that "there is no practically possible way of observing the neutrino." Bethe responded with good humor: "Well, you shouldn't believe everything you read in the papers."

Through their painstaking efforts at Hanford and Savannah River, Reines and Cowan succeeded in trapping the elusive beast for the first time. They captured the ghost particle that can pass through the Earth unhindered and travel to the far reaches of the universe, and they did so with an ingenious experimental setup using a nuclear reactor as a source, just as Bruno Pontecorvo had proposed, but using a different detection technique from what Pontecorvo had in mind. The phantom conjured up by Pauli's tortured psyche became a tangible constituent of our material world, providing us with a satisfactory solution to the mystery of beta decay and helping us preserve the law of energy conservation, thus validating the theoretical predictions made by Pauli and Fermi.

Since then, we have realized that the shy neutrinos hold the key to unraveling a great many cosmic mysteries, from what makes the Sun shine to why the universe is not completely empty of matter. The farsighted Pontecorvo was the first to suggest that the Sun should emit tremendous numbers of neutrinos, and next we turn to the story of a trailblazer who was determined to catch these messengers from afar.

FOUR

SUN UNDERGROUND

Fred Reines and Clyde Cowan were not the only scientists chasing neutrinos in the early 1950s. Ray Davis, a Yale-trained physical chemist, caught neutrino fever at a library and was hot on their trail too. Growing up in Washington, D.C., Davis played street games and paddled on the Potomac River with his younger brother, just fourteen months his junior. His father, a photographer at the National Bureau of Standards who had never finished high school, encouraged Davis's interest in chemistry experiments and photography. Young Davis was also good at rifle shooting and won medals for his marksmanship, though he later swore off the activity. His mother taught him to enjoy music, but unlike Reines, Davis was a poor singer. According to Davis, "It was to please [my mother] that I spent several years as a choirboy, in spite of my inability to carry a tune."

As he grew older, Davis decided to go into science while his brother pursued a military career. After completing a chemistry doctorate at Yale, Davis joined the army as a reserve officer and was employed as an observer of chemical weapons tests in Utah during the Second World War. He spent his spare time exploring the surrounding area and taking photographs. Later he worked for the Monsanto Chemical Company, investigating radioactive materials, before he joined the newly created Brookhaven National Laboratory in 1948. The laboratory, built on a surplus army base on Long Island after the war, was tasked with exploring peaceful applications of nuclear science. It was at Brookhaven that Davis met his wife, who worked in the laboratory's biology department. They had five children over the next fifteen years. Living in a seaside community, he built his own sloop, with his wife's help, and took up what became a lifelong interest in sailing.

When Davis first arrived at the laboratory, he asked his boss what he should do with his time there. As Davis recalled decades later, "To my surprise and delight, he told me to go to the library and find something interesting to work on." That's how he came across an article on neutrinos. Reading it, Davis realized that scientists knew little about these mysterious particles, despite the pioneering theoretical work of Wolfgang Pauli, Enrico Fermi, and Bruno Pontecorvo, so the topic was wide open for experimental investigations. What intrigued him most was a description of Pontecorvo's proposal for detecting neutrinos using a big tank of chlorine. Pontecorvo had noted that if a neutrino interacted with a chlorine atom, it would turn into a radioactive argon atom, which could be identified when it decayed and emitted radiation. Given Davis's training in the chemistry of radioactive materials, he was ready to take on the challenge. That day at the Brookhaven library, Davis found his life's calling, one that he doggedly pursued against long odds.

Davis worked on a number of other scientific questions over the years, and was particularly keen to measure radioactive elements in meteorites in order to infer their ages and life histories. Together with a colleague, he applied radiometric dating techniques—which rely on comparing the abundances of radioactive isotopes and their decay products to measure time—to determine how long these rocks had been floating around in space, exposed to cosmic rays, before they landed on Earth. When Apollo astronauts brought back Moon rocks, Davis was among the researchers who got to analyze their composition, which resulted in an amusing incident. As he recalled, "During processing of the *Apollo 12* samples, one of the glove boxes in Houston leaked and I had the interesting experience of being quarantined with the astronauts and a few other unlucky scientists for two weeks until it was clear that we were not infected with any lunar diseases." Despite his forays into other research areas, Davis was most interested in chasing neutrinos throughout his career.

For his first attempt at neutrino hunting, Davis set up a 3,800-liter (1,000-gallon) tank of dry-cleaning fluid (or carbon tetrachloride) next to a modest nuclear reactor at Brookhaven itself. He knew that neutrinos rarely interacted with matter, so he waited several weeks, hoping that would allow enough time for a handful of reactions to occur, and measured the amount of argon that accumulated. The results were disappointing: there was no extra argon beyond what cosmic ray collisions could account for. There was no sign of neutrinos. Davis tried again in 1955, this time building a bigger version of his apparatus next to the much more powerful reactor at the Savannah River Site in South Carolina, the same place where Reines and Cowan conducted their experiment. But again, Davis found nothing. While Davis came up empty-handed, Reines and Cowan pinned down the ghostly particle by the following year, using scintillator liquid and photomultiplier tubes instead of the chlorine-argon method suggested by Pontecorvo. But the game was far from over for Davis. Now that Reines and Cowan had confirmed these particles as real, Davis turned his attention to chasing neutrinos created inside the Sun rather than those produced in man-made reactors on Earth.

Davis was aware that neutrinos should be an essential by-product of how the Sun generates energy, thanks to astrophysicists who had figured out the Sun's inner workings in the preceding decades. The first crucial insight on solar energy production came in 1920 from the British astronomer Arthur Eddington, who suggested that nuclear reactions are responsible. One of Eddington's Cambridge colleagues had found that a helium atom is slightly less massive than four hydrogen atoms combined. Eddington proposed that as four hydrogen nuclei fuse to make one helium nucleus in the Sun's core, the small mass difference would be converted into energy, according to Einstein's equation $E=mc^2$. It was a stroke of brilliance on Eddington's part, but he didn't work out the details of the mechanism. He left it for someone else to figure out a nuclear chain reaction that was fast enough to account for the Sun's brilliance but not so fast that the Sun should have burned out long ago.

Hans Bethe, a versatile theorist at Cornell University in Ithaca, New York, took on the challenge of understanding the solar nuclear engine. Bethe was born in 1906 in Strasbourg, which at the time belonged to the German Empire but is now part of France. His father was a physician and his mother was an accomplished musician, until she lost much of her hearing after contracting influenza. Perhaps as a result of the illness, she suffered from bouts of depression, and Bethe's parents eventually got divorced. Bethe was fascinated by numbers from about age four, and taught himself calculus by fourteen. He also learned to write early and filled little books with stories, but he had a peculiar habit as a child: he wrote one line from left to right and the next line from right to left (like the Greeks of the seventh century B.C.E.). By the time Bethe finished high school, he was more interested in physics because "mathematics seemed to prove things that are obvious." After two years at Frankfurt University, he moved to Munich to study under the charismatic physicist Arnold Sommerfeld, just as Wolfgang Pauli had several years earlier.

Bethe proved to be exceptionally talented at theoretical physics, completing a doctorate with the highest honors. He spent part of the following year in Rome, working with Fermi, whom he admired greatly. In a letter to his former thesis adviser Sommerfeld, Bethe wrote, "The best thing in Rome is unquestionably Fermi. It is absolutely fabulous how he immediately sees the solution to every problem that is put to him." Fermi taught Bethe how to derive quick insights from back-of-the-envelope estimates, a much less formal approach to physics than he had learned in Munich under Sommerfeld.

Bethe returned to Germany in 1932 to take up a teaching position, but lost it a year later because of his mother's ancestry when Adolf Hitler introduced racial laws barring Jews from holding government positions. Like many other European scientists of Jewish descent, Bethe soon ended up in America, in his case as a professor at Cornell, where he enjoyed the collaborative scientific atmosphere and continued his work on nuclear physics. One of his actions did provoke the ire of many colleagues: he broke off his engagement to Hilde Levi, whom he had met in his Frankfurt days and grown close to over the years, because of his mother's vehement objections, just days before the wedding. Levi's friend the great Danish physicist Niels Bohr was so dismayed that he shunned Bethe for years.

Early in 1938 Bethe and Charles Critchfield, a graduate student at George Washington University at the time, studied the series of nuclear reactions that we now call the "p-p chain" (p-p stands for proton-proton), one of the ways stars convert hydrogen to helium and release energy. Bethe and Critchfield calculated the rate at which two protons fuse together, overcoming mutual electrical repulsion, to make a deuteron, a loosely bound nucleus consisting of a proton and a neutron. As one of the original protons turns into a neutron, it would emit a positron and a neutrino. The researchers posited that the deuteron would capture another proton quickly, and turn into a nucleus of helium-3. In the final step, two helium-3 nuclei would form a stable helium-4 nucleus and release two protons. In a nutshell, Bethe and Critchfield found a nuclear chain reaction that turns four protons (or hydrogen nuclei) into one helium-4 nucleus, while emitting photons (energy), positrons, and neutrinos. In their picture, positrons and electrons would annihilate each other and produce energetic gamma rays. These gamma ray photons would bounce their way outward through the Sun's various layers, taking hundreds of thousands of years to reach the surface. By then the photons would have lost a lot of energy, so they would emerge as visible light. Neutrinos, on the other hand, should escape unhindered and reach the Earth in just over eight minutes, traveling at nearly the speed of light.

However, Bethe and Critchfield did not have a good estimate of the temperature at the solar core. So the energy production rate they calculated turned out to be much higher than the Sun's actual output. But in March 1938, Bethe heard encouraging news at a meeting in Washington, where astrophysicists presented new, lower temperature estimates for the Sun's interior. He realized that these revised values would bring his

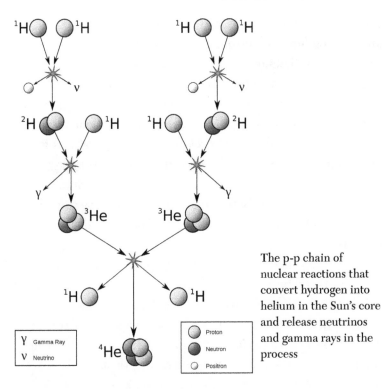

The p-p chain of nuclear reactions that convert hydrogen into helium in the Sun's core and release neutrinos and gamma rays in the process

calculations into better agreement with observations, and decided to examine all the different routes for converting hydrogen into helium inside stars.

Using just pen and paper, Bethe discovered an alternate chain reaction that we now call the CNO cycle—"CNO" standing for carbon, nitrogen, and oxygen—that would also fit the bill. As Bethe recalled, "I did not, contrary to legend, figure out the carbon cycle on the train home from Washington. I did, however, start thinking about energy production in massive stars upon my return to Ithaca." Within two weeks of the Washington meeting, he worked out details of the cycle, which began with a carbon atom absorbing a series of protons (or hydrogen nuclei) and turning first into nitrogen, then into oxygen. The oxygen in turn emitted a helium nucleus and converted back to carbon. It was an elegant way to turn hydrogen into helium while releasing energy, using carbon as a catalyst. The nuclear reactions that constituted the CNO cycle also produced neutrinos. The problem was that the CNO cycle needed temperatures above 20 million degrees to work. So it described the energy production in stars more massive, and hotter, than the Sun, but not in the Sun itself. Thus Bethe concluded that massive stars shone via the CNO cycle, while the Sun itself probably depended on the p-p chain.

If Bethe's theory for solar energy production was correct, the Sun should be a prodigious source of neutrinos. In his 1939 paper titled "Energy Production in Stars," Bethe made no mention of detecting neutrinos as a way to test the theory, however. These elusive particles remained a theoretical figment at the time, so Bethe's omission was not surprising. In fact, solar neutrinos got only a passing mention even in Pontecorvo's prophetic 1946 report, but the prospect of probing the Sun's heart through neutrinos caught Ray Davis's fancy.

In fact, Davis used his experiment at Brookhaven to search for solar neutrinos. His apparatus was nowhere near sensitive enough to detect them, but he placed an upper limit on the number of solar neutrinos arriving each second and wrote up his findings for publication. One scientist was outright dismissive of Davis's upper limit, declaring, "One would not write a scientific paper describing an experiment in which an experimenter stood on a mountain and reached for the moon, and concluded that the moon was more than eight feet from the top of the mountain." Such skeptics didn't deter the plucky experimentalist. Davis's first attempt was a small step, for sure, but an important one nonetheless.

The biggest obstacle that Davis faced was that most neutrinos produced during the p-p chain were too low in energy to interact with chlorine to produce argon, and so they were undetectable with his experiment. But Davis still had hope, because not all neutrinos were created equal, and some should be energetic enough for detection. In particular, he knew that every so often the third step of the p-p chain would proceed differently: instead of two helium-3 nuclei combining to form helium-4, a helium-3 would fuse with a helium-4 to make beryllium-7. In turn, beryllium-7 could react with a proton to become boron-8, which is unstable and would decay to beryllium-8 by emitting a positron and a neutrino. *This* neutrino would have enough energy to react with chlorine in Davis's experiment. Luckily for Davis, in 1958 two physicists at the Naval Research Laboratory in Washington discovered that this alternate reaction occurred a thousand times more frequently than researchers had previously suspected. Two astrophysicists, Willy Fowler at Caltech and Alastair Cameron, then at Chalk River Laboratories in Canada, realized what this discovery meant for solar neutrino hunting and alerted Davis to the improved prospects for success.

Encouraged by the news, Davis decided to search for solar neutrinos again in late 1959. This time he took his experiment underground, to the 2,300-foot-deep Barberton limestone mine in Ohio, to get away from pesky cosmic rays that might otherwise drown out the neutrino signal. His early estimates for the sensitivity of the experiment were rather optimistic: he believed it should register several neutrinos from the Sun each day. But Davis was in for disappointment once more: when he examined the results, he didn't find any sign of the elusive messengers from the Sun. Soon after, Davis received more bad news. While researchers at the Naval Research Laboratory had shown that beryllium-7 was easy enough to make, other scientists found that the next crucial step, when it absorbs a proton to form boron-8, was difficult. That meant the rate of high-energy neutrinos arriving from the Sun would be too low for Davis's experiment to register. As Fred Reines summed up in 1960, "The probability of a negative result even with detectors of thousands or possibly hundreds of thousands of gallons of [carbon tetrachloride] tends to dissuade experimentalists from making the attempt." The situation seemed hopeless to most physicists. Someone less stubborn than Davis might have cut his losses and moved on. Instead, Davis considered scaling up his experiment to be a hundred times bigger—about the volume of an Olympic-size pool—and thus much more sensitive.

Meanwhile, two other scientists, who made contact with each other through a lucky accident, intervened to change the course of events. It's interesting how often chance occurrences, like the grounding of a plane by engine trouble that brought Fred Reines and Clyde Cowan together, affect the development of a scientific discipline. In this case, a journal editor served as the unwitting intermediary between two scientists who hadn't known each other previously. One was Davis's friend Willy Fowler, who had been instrumental in showing that nuclear processes inside stars were responsible for making the full range of elements from carbon to iron, starting with just hydrogen and helium produced in the big bang. The other was a bright young theorist by the name of John Bahcall. Growing up in Louisiana, Bahcall was a fine tennis player and a champion debater in high school. His ambition was to study philosophy and become a rabbi. After a year at Louisiana State University, he went to the University of California, Berkeley, to take a summer course, loved it there, and stayed for an undergraduate degree in philosophy, thanks to a relative who agreed to cover his tuition.

Before he could graduate, he was required to take a science course. So he persuaded a professor to let him take a class for physics majors, even though he had never taken science in high school. That's when Bahcall discovered a passion for physics. As he recalled later, "It was the hardest thing I have ever done in my life, but I fell in love with science. I was thrilled by the fact that by knowing some physics you could figure out how real things worked, like sunsets and airplanes, and that after a while everyone agreed on what was the right answer to a question." Bahcall went on to earn a master's degree in physics at the University of Chicago and a PhD at Harvard.

In 1960, while he was a research fellow at Indiana University, Bahcall submitted a paper on beta decay processes in stars to the *Physical Review*. To his surprise, even before the paper appeared in print, he received a letter about it from Willy Fowler, who had been asked by the journal editor to referee the article—and with the letter came an offer to work at Caltech. Fowler was so impressed with Bahcall's work that he also wrote to Ray Davis about the young theorist, and encouraged Davis to contact him. So Davis wrote to Bahcall, asking for his help to improve the predictions for solar neutrino production by calculating the rates of relevant nuclear processes. Bahcall was happy to oblige, and so began a close scientific collaboration and a personal friendship that lasted for over five decades.

At first Bahcall underestimated the difficulty of the task that Davis had suggested. As Bahcall explained in an interview for *Nova* decades later, "When I got to Caltech and began the calculations of how many neutrinos there should be from the Sun, I realized the problem was immensely more complicated than I had recognized early on, because there were many different reactions competing with each other." What's more, he needed to determine the characteristics of the Sun's interior—such as chemical composition, temperature, density, and pressure—to high accuracy before he could provide Davis with robust estimates of solar neutrino production. The results of Bahcall's first calculations were not encouraging: he predicted that Davis's 3,800-liter tank would register only one neutrino every hundred days. According to Bahcall's estimates, even if Davis were to build a tank with a hundred-times-larger volume, he would capture only one particle daily, far from the numbers needed to make a reliable measurement of the Sun's nuclear engine.

But better news came from another front. Following up on a Danish physicist's suggestion, Bahcall found that chlorine should capture solar neutrinos twenty times more efficiently than he had previously thought. This provided an enormous boost to Davis's plans. Now, with the prospect of catching a handful of solar neutrinos every day, Davis saw a good reason to build a bigger detector. He knew that in order to shield such a sensitive experiment from cosmic rays, he needed to place it a mile underground.

Even though there was no funding for his grand scheme, Davis began to look for suitable deep mines in the United States with the help of his Brookhaven colleague Blair Munhofen in 1963. A Bureau of Mines official recommended two possibilities: the Homestake Gold Mine in South Dakota and the Anaconda Copper Mine in Montana. Davis and Munhofen visited both sites to judge for themselves. The owners of the Anaconda mine were eager for the scientists to choose their site, and offered to provide concrete lining for a cylindrical hole at low cost. But the Brookhaven scientists reckoned the rock at the Anaconda would not permit excavating a large enough cavity at the right depth. At the Homestake, on the other hand, opening a large cavity was no problem, but the cost estimate for the excavation came in high. So the two scientists decided to look for other options. Finally, they identified the Sunshine Mine, a silver mine in Idaho, as a suitable location with a reasonable price tag. If only they could find the money to build the experiment, now there was a place to put it.

It was time to get serious. Davis and Bahcall decided to approach Brookhaven's director, Maurice Goldhaber, for formal approval and funding. Goldhaber was a nuclear physicist who doubted the ability of astronomers to calculate anything with sufficient precision to be of interest to him. Davis knew of Goldhaber's bias, and advised Bahcall to focus on nuclear physics rather than astrophysics during their discussions with the director. The tactic worked, and Goldhaber gave his backing and funding from the laboratory budget. Bahcall later described the feat admiringly as "Ray's greatest political achievement."

In early 1964 Bahcall and Davis outlined their theory and the experiment in back-to-back papers, making the case for a 380,000-liter (100,000-gallon) tank of dry-cleaning fluid to hunt for solar neutrinos. They attracted quite a bit of attention in the broader scientific community. From his new home in Russia, Bruno Pontecorvo read the papers with avid interest. Even the popular press covered the pair's plans to hunt for solar neutrinos, and the publicity resulted in unexpected benefits: when plans for the Sunshine Mine fell through (the aptness of its name notwithstanding), the owners of the Homestake mine came up with a much more favorable cost estimate for excavation, and tank manufacturers were more interested in helping with the experiment. As Davis noted in a letter to Bahcall around this time, "These tank people take us more seriously after the article in *Time*."

The excavation of the rock at the Homestake mine began in the spring of 1965 and took about two months, and when Davis and Blair Munhofen went to examine the enormous cavity almost a mile belowground, they were pleased with what they saw. Meanwhile, Davis hired the Chicago Bridge and Iron Company, which had experience in making leakproof space chambers for NASA, to build the tank. The company workers completed the vessel within a year, cleaned it thoroughly, and sealed it to prevent atmospheric argon from leaking in, to avoid possible contamination. Davis and Bahcall later learned that the company ordinarily "would not have been interested in building a small, rather conventional tank such as was required for the neutrino experiment but they were intrigued by the aims of the project and the unusual location." Next, ten railroad cars full of dry-cleaning fluid arrived at the mine from Kansas. The scientists hauled the liquid down to the tank in specially designed containers, using the mine's rail system and a hoist. The final step of the preparation was to purge any air that was dissolved in the fluid, in order to remove any traces of argon.

By the fall of 1966, the experiment was ready to begin. The final price tag came to $600,000, or as Davis put it, "ten minutes' time on commercial television." In the meantime, Bahcall had continued to refine his calculations of how many solar neutrinos Davis should detect with the new apparatus. According to Bahcall's best estimates, neutrinos interacting with chlorine

Construction of the tank that Ray Davis used for his solar neutrino experiment in the Homestake Gold Mine (Brookhaven National Laboratory)

should produce a few dozen argon atoms every few weeks. Davis was confident he could fish out almost every one of them. He wasn't inclined to make grandiose proclamations, however, and described his work modestly as "just plumbing," meaning that the all-important hunt for solar neutrinos came down to the mundane task of building and operating the detector tank and the associated pipes, taking extreme care to leave no room for leaks. But, as Bahcall wrote, "as a nonchemist I am awed by the magnitude of his task and the accuracy with which he can accomplish it . . . He is able to find and extract from the tank the few dozen atoms of [radioactive argon] that may be produced inside by the capture of solar neutrinos. This makes looking for a needle in a haystack seem easy."

In fact, Davis had to go through several intricate steps to pick out his quarry. After waiting several weeks for neutrino interactions to produce some argon atoms, he flushed out the tank with helium gas, which carried the argon along with it to a cooled charcoal trap. At a very low temperature, the argon condensed out into the trap, separating from the helium. Later Davis heated the trap to release the argon as a gas, collected it, and purified it chemically to remove any traces of other radioactive elements. The final gas sample, about the size of a small sugar cube, would contain normal argon as well as a handful of argon-37 atoms produced by neutrino interactions. Davis used a Geiger counter to measure the number of radioactive argon atoms, which in turn revealed the number of high-energy neutrinos coming from the Sun. Culling just a few specific atoms out of a tank that contained about one million trillion trillion (or 10^{30}) atoms altogether was an astonishing achievement indeed.

After two years of data gathering at the Homestake mine, Davis announced the first results from his experiment at a meeting at Caltech in 1968. He claimed to detect solar neutrinos, but only about one-third as many as Bahcall's model calculations predicted. To detect any solar neutrinos at all, to peer into the heart of a star for the first time, would have been a remarkable feat—but what made headlines was the large discrepancy between theory and observation.

Bahcall feared that Davis's results meant his solar model was wrong. The young theorist appeared so glum at the Caltech meeting that the legendary physicist Richard Feynman, who had received a share of the physics Nobel Prize three years earlier for his work on the theory of quantum electrodynamics, asked him whether he would like to go for a walk. The two strolled around the campus, making small talk. Eventually, as recalled by Bahcall, Feynman tried to console him: "Look, I saw that after this talk you were depressed, and I just wanted to tell you that I don't think you have any reason to be depressed. We've heard what you did, and nobody's found anything wrong with your calculations. I don't know why Davis's result doesn't agree with your calculations, but you shouldn't be discouraged, because maybe you've done something important, we don't know." Feynman's kind gesture and encouraging words touched Bahcall and helped lift his spirits.

The mismatch between theory and data raised doubts not only about Bahcall's model but also about the reliability of Davis's experiment. Many scientists were skeptical that Davis had detected solar neutrinos at all. They wondered whether some outside air, which contains argon, had leaked in and contaminated the fluid. Besides, how reliably could Davis flush out a few argon atoms from a giant volume of dry-cleaning fluid? As a way to address the criticism, Willy Fowler dared Davis to put his methods to the test by injecting 500 atoms of radioactive argon into the fluid, mixing it up, and retrieving them all. Davis met the challenge with ease: he extracted every one of the argon atoms.

The test results convinced some doubters that Davis's experimental technique was sound. Now they questioned Bahcall's solar model. Perhaps Goldhaber, the Brookhaven director, was right all along: astrophysicists didn't really know what they were talking about. Others wondered if the findings were a statistical fluke. Given the small number of events involved, that was a possibility, just as a coin tossed a few times could come up heads each time by chance. In an effort to address these concerns and improve reliability, Davis tinkered with the detector to discriminate better between true neutrino events and unrelated background "noise." Bahcall refined his solar model, incorporating new laboratory measurements of relevant nuclear reaction rates into his calculations. Unfortunately, none of these improvements, nor years of gathering more data, changed the bottom line: the big disparity between the predictions and the observations didn't disappear.

Despite the diligent work of Davis and Bahcall, by the early 1970s it was clear that the "solar neutrino problem" was not going away anytime soon. That prompted many scientists to come up with possible solutions, ranging from the reasonable to the outlandish. Several people suggested tweaks to the standard solar model such as changing the abundance of heavy elements, speeding up the rotation rate of the core, or adding the effects of a magnetic field. The Australian mathematician Andrew Prentice made a more dramatic, if not terrifying, proposal: he suggested that the Sun had already burned out, leaving a helium core. Since it takes tens of thousands of years for the photons to emerge from the solar core, he reasoned, it would be a while before we would find out. The British astrophysicist Fred Hoyle, known for his contrarian views, thought that the solar core was made mostly of heavy elements, surrounded by a hydrogen envelope. Another group of theorists proposed that there was a black hole at the center of the Sun: in their view, instead of nuclear fusion, it was energy released by matter falling into the black hole that powered the Sun. Others thought the problem had to do with our misunderstanding of neutrino properties. Bahcall himself considered whether neutrinos might be unstable and decay into other particles. Pontecorvo and a Russian colleague proposed that neutrinos come in multiple flavors and switch among them on their way from the Sun to the Earth. Davis's detector, they argued, registered only one flavor of neutrinos, which they believed would explain the shortfall.

By the start of the 1980s, the Homestake experiment had passed the threshold of credibility in the minds of most physicists. Davis's results were no longer in dispute. It was clear he had made painstaking efforts to understand and refine his apparatus, yet the annoying discrepancy remained unaccounted for. "It is surprising to us, and perhaps more than a little disappointing, to realize that there has been very little quantitative change in either the observations or the standard theory since these [first] papers appeared [in 1968], despite a dozen years of reexamination and continuous effort to improve details," Bahcall and Davis wrote. Still, Davis's was the only experiment that was hunting these ghostly messengers from the Sun, and few researchers wanted to delve into a field that appeared to have reached an impasse. Bahcall summed up the situation best: "All the people working steadily on solar neutrinos (theorists and experimentalists) could (and often did) fit comfortably into the front seat of Ray's car." Meanwhile, the mismatch between the predictions and the detections of solar neutrinos remained, as a *New York Times* writer described it, "one of the biggest embarrassments of twentieth-century science."

Most scientists lay the blame for the disparity at the feet of astronomers. The problem, they argued, was with Bahcall's model of the solar interior. But Bahcall found a new reason to be optimistic by the late 1980s: the first measurements of how the Sun vibrates, from the emerging field of helioseismology, were in agreement with his model predictions. Feeling vindicated, Bahcall announced it was time for astronomers to declare victory: the helioseismology results meant the solar models must be correct, and could not be responsible for the neutrino discrepancy. Many physicists disagreed with his assessment. The summary speaker at one conference made fun of Bahcall's claims, using humorous caricatures as viewgraphs. As Bahcall reminisced, "He had the whole auditorium, including me, laughing at the bravado, the hubris of this guy claiming that he could say something about particle physics based on this complicated Sun." After this humiliation in front of his colleagues, Bahcall toned down his conclusions.

Meanwhile, scientists at a new experiment on the other side of the world, in the Kamioka mine in Japan, some 150 miles west of Tokyo, joined the hunt for solar neutrinos in the mid-1980s. The original goal of their detector was to look for the possible decay of the proton; hence it was named Kamiokande for Kamioka *n*ucleon *d*ecay *e*xperiment. The experiment was the brainchild of Masatoshi Koshiba, who had returned to his home country after working in the United States for several years. He and his colleagues wanted to test a key prediction of the so-called grand unified theories.

The GUTs, as these theories were popularly dubbed, attempted to provide a single framework to describe three of the four fundamental forces of nature. Theorists suggested that even though the electromagnetic, weak, and strong forces appeared distinct and different from each other in the present-day universe, they had acted as a single unified force soon after the big bang. If these theorists were right, one consequence would be that protons might disintegrate spontaneously into lighter particles, but do so over extremely long timescales. Koshiba knew that even if the average lifetime of a proton were much longer than the age of the universe, he should be able to catch a few decays every year if he could monitor an enormous number of protons. So he persuaded his colleagues to build the Kamiokande detector, which consisted of a tank containing 3,000 tons of pure water and a thousand phototubes mounted on its inside walls. They didn't find any evidence of proton decay, however.

But the researchers soon realized that they could use their apparatus to measure neutrinos from the Sun instead. So they modified and upgraded their experiment, with the help of some American colleagues, to make it sensitive to solar neutrinos. In searching for solar neutrinos, the Kamiokande worked rather differently from the one at the Homestake mine. It used water instead of dry-cleaning fluid to trap the particles. Every once in a while, a solar neutrino would collide with an electron in the water and propel it forward, like a billiard ball that's hit head-on. The fast-moving electron would create an electromagnetic "wake," or a cone of light, along its path. The resulting pale blue light is called "Cherenkov radiation," after the Russian physicist Pavel Cherenkov, who investigated the phenomenon. Photo-tubes lining the inside walls of the tank would register each light flash and reveal an electron's interaction with a neutrino. The Kamiokande provided two extra bits of information to re-searchers: from the direction of the light cone scientists could infer the direction of the incoming neutrino, and from its inten-sity they could determine the neutrino's energy. On the down-side, the Kamiokande detector was only sensitive to high-energy neutrinos, just like the Homestake detector. The other disad-vantage was that phototubes registered other signals, such as cosmic ray hits, in addition to neutrino events. But the research-ers figured out how to distinguish true neutrino events from other sources of noise from the pattern of light.

The Kamiokande team reported the early results of their solar neutrino hunt in the summer of 1989. Their independent findings vindicated Ray Davis: not only did the Kamiokande detector confirm that neutrinos came from the direction of the Sun but it also found a deficit in the particle number predicted by Bahcall, just as the Homestake detector had. And yet, over the next few years, the Kamiokande team also confirmed that the energy spectrum of the incoming neutrinos agreed with Bahcall's calculations. Thus it appeared that Davis and Bahcall were both right, and the solar neutrino deficit was real. So what gives? According to Bahcall, who found great comfort in the Kamiokande results, "My feeling was, aha, we've eliminated the possibility of experimental results being wrong, and I'm confident in my theory. I think we're onto something good." As it turned out, the maddening discrepancy led to new physics—though not new for Bruno Pontecorvo, who had pointed the way decades earlier.

FIVE

COSMIC CHAMELEONS

Bruno Pontecorvo had two crucial insights over a half century ago that contained the keys to solving the solar neutrino puzzle. His first insight was that there was more than one variety of neutrinos. He came to this conclusion while examining the decay of an unstable particle called a muon, which belongs to the lepton family, along with the electron and the neutrino. Leptons do not take part in the strong interaction and are fundamental building blocks of matter. The muon is also negatively charged, but about two hundred times more massive than the electron, and it lives for just over two-millionths of a second before breaking up. Pontecorvo proposed that the muon and the electron each had a distinct variety of neutrino associated with it.

Three physicists at Columbia University—Leon Lederman, Melvin Schwartz, and Jack Steinberger—confirmed the existence of two neutrino varieties while experimenting with a particle collider in 1962, and proved Pontecorvo right on this score. When Martin Perl of Stanford University and his colleagues identified a third, even more massive, member of the lepton family, called the tau particle, researchers expected that there should be a third type of neutrino associated with it. Physicists at Fermilab near Chicago finally observed tau neutrinos in the year 2000. We use the whimsical term "flavors" to describe the three neutrino types.

Pontecorvo's second insight was that neutrinos could be fickle. He found that the laws of quantum mechanics allowed neutrinos to morph, or "oscillate," between types, but this could only happen if they had some mass. He realized that the neutrino's mass could be minuscule, even thousands of times smaller than that of the electron, but it could not be zero. He also knew that nuclear reactions in the Sun produced only one flavor of neutrinos—electron neutrinos—and that it was the only variety that Ray Davis's chlorine-based experiment could detect. Soon after Davis first reported a deficit of solar neutrinos in 1968, Pontecorvo and his Russian colleague Vladimir Gribov proposed that neutrinos oscillating from one flavor to another on their way from the Sun could account for the shortfall. It was as if they had suggested that chocolate ice cream could turn into vanilla, but as weird as the theory may sound, their suggestion offered a simple and elegant explanation for the missing solar neutrinos: two-thirds of the electron neutrinos produced in the Sun could turn into other varieties during their long journey, and thus escape detection.

Their proposal makes perfect sense in the wacky world of quantum mechanics, where certainty gives way to probability. In quantum mechanics we can also describe particles in terms of waves, whose wavelength depends on the mass and the speed of the particle. Mathematically speaking, there is a "wave function" that describes each flavor of neutrinos. If each of the three neutrino flavors had a different mass, the three wavelengths would also be different. In fact, what we think of as a

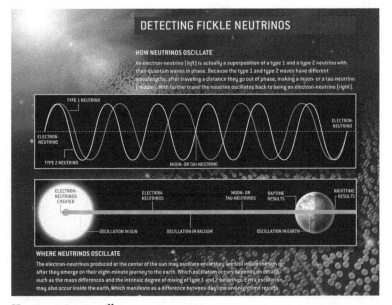

How neutrinos oscillate

neutrino particle is a hybrid of all three flavors. In the ice cream analogy, it is as if we had swirls of chocolate, vanilla, and strawberry flavors blending together. As a neutrino travels through space, the waves associated with the three flavors advance at different rates. The waves mingle with each other along the way, so at different points in space you get a different mixture of flavors. Sometimes you would taste mostly chocolate, while at other times vanilla or strawberry would dominate. So, a neutrino that was born as an electron neutrino could appear as a tau neutrino after some distance. That's why, as Pontecorvo and Gribov suggested, neutrinos could oscillate between flavors as they travel from the Sun to the Earth.

The problem was that their suggestion ran counter to the

conventional wisdom among physicists, most of whom assumed that neutrinos were massless and traveled at the speed of light, just like photons, which also lacked mass. If that were the case, these particles wouldn't be permitted to oscillate between flavors. In fact, the standard model of particle physics, formulated in the 1970s and vindicated by many experiments since, assigned a zero mass to neutrinos. Given the standard model's spectacular success at describing the subatomic realm, few physicists were willing to abandon its tenet regarding neutrinos and embrace Pontecorvo's radical proposition.

However, their sentiments began to change when three theorists stumbled upon an intriguing possibility in their calculations. Stanislav Mikheyev and Alexei Smirnov in Russia, building on an insight of Lincoln Wolfenstein in the United States, realized that the presence of matter would enhance neutrino oscillations a great deal. So if neutrinos produced in the solar core started off with a mild tendency to swap identities, they would end up totally schizophrenic by the time they reached the Sun's surface. Many physicists found the mathematical description of this phenomenon, called the MSW effect after the initials of its discoverers, rather compelling. As John Bahcall told a journalist, "The MSW effect is a beautiful idea. It would seem like a cosmic mistake if nature did not use this solution."

But theoretical considerations alone weren't sufficient to convince physicists that Pontecorvo had been right about neutrino oscillations. So, many researchers were excited when clearcut experimental evidence of neutrinos morphing between flavors came to light in the 1990s. By then, Japanese neutrino hunters had upgraded their detector, now called Super-Kamiokande or Super-K, to be many times more sensitive than the original Kamiokande detector. Super-K, like its predecessor, could record not only solar neutrinos but also neutrinos produced by cosmic rays hitting the Earth's upper atmosphere. These so-called atmospheric neutrinos are hundreds or even thousands of times more energetic than those coming from the Sun, so they are easier to trap. Scientists estimated that muon neutrinos should be twice as common as electron neutrinos among the cosmic ray debris. Fortunately, the Super-K detector was able to distinguish between these two neutrino types: an electron neutrino hitting the water would produce a fuzzy circle of light, whereas a muon neutrino interaction would lead to a sharp ring. After observing atmospheric neutrinos of both types for nearly two years, the Super-K team reported a surprising result: instead of twice as many of the muon variety, they found roughly equal numbers of the two types. One possibility, they reasoned, was that half the muon neutrinos were morphing into the third type, tau neutrinos, which Super-K could not identify easily.

The most intriguing clue had to do with the direction from which neutrinos arrived. Roughly equal numbers of cosmic rays should hit the Earth's atmosphere from all directions, so the number of neutrinos produced by these particle collisions should also be the same all around the globe. Sure enough, the Super-K researchers found equal numbers of electron neutrinos coming down from the sky and coming up through the ground, from the other side of the Earth. But that wasn't true for muon neutrinos: only half as many were coming up from below as coming down from overhead. It seemed to the Super-K team that muon neutrinos were somehow disappearing during their journey through the Earth. "That was the smoking gun," as Ed Kearns of Boston University, a member of the Super-K collaboration, put it. Most likely, they concluded, the muon neutrinos were changing identity, morphing into tau neutrinos that Super-K couldn't detect readily. Thanks to these findings, by the late 1990s many more physicists were willing to accept that oscillating neutrinos could be responsible for the atmospheric neutrino anomaly as well as for the solar neutrino deficit.

However, showing that some muon neutrinos disappear midflight wasn't direct proof of their metamorphosis into a different variety. To be sure this interpretation was correct, physicists needed to measure what the electron neutrinos from the Sun turned into, or at least measure the electron neutrinos separately from the other flavors. That was the primary goal of the Sudbury Neutrino Observatory (SNO), which was built inside an active nickel mine in northern Ontario in order to solve the solar neutrino riddle once and for all. The laboratory has since expanded into a larger facility, called SNOLAB.

Its director, Nigel Smith, agreed to give me a tour, so one late November afternoon in 2010, I drove north for four hours, through intermittent snow flurries, from Toronto to Sudbury. The next morning, in the predawn darkness without a GPS device to depend on, I nearly got lost driving from the B&B where I had stayed to the SNOLAB site, but managed to arrive just in time to catch the last elevator that went down at 7:00 a.m.

Inside a locker room at the ground level, donning blue overalls and steel-toed boots, Nigel Smith fastened a light on his hard hat and a battery pack on his safety belt, and asked me to do the same. After placing two tags—one for Smith and the other for a "visitor"—on a peg wall so that it would be easier to take a tally in case of an emergency, we stepped into a dark, creaky elevator suspended by a cable almost as thick as my arm. Two dozen miners packed into the open cage with us. Our drop down to the pits of the Earth began slowly, but soon picked up speed. The headlamps provided just enough light for me to make out the rocky walls of the mine shaft rushing past in front of us. The cage made several stops on its way down to let out groups of miners, and I caught glimpses of lighted tunnels receding into the distance at each level. About halfway down, my eardrums could feel the pressure change, so I worked my jaws and forced a yawn. At the final stop, just over a mile and a quarter below the surface, Smith and I stepped out, along with the few remaining miners. Our descent, including the stops along the way, had taken about ten minutes.

Our journey was far from over, however, since we still had more than a mile-long trek through a muddy tunnel ahead of us to reach SNOLAB. Thankfully, a combination of concrete props, roof bolts, and steel screens held off the rock overhead from crumbling under pressure, and a ventilation system produced a cool breeze, without which we'd be sweating buckets. The miners veered off to side tunnels in search of nickel (the primary reason for excavating the mine), while Smith and I kept on going straight, walking along rail tracks laid for trolleys. At last we reached a sign that declared SNOLAB: MINING FOR KNOWL-EDGE, signaling that we had arrived at the world's deepest underground laboratory. We washed the mud off our boots with a hose and pulled open a bright-blue door. I was immediately struck by the contrast between the pristine laboratory compound inside, with spotless floors, shiny walls, and dust-free air, and the grimy mine we had just walked through. Before going farther, we took showers and changed into a new set of overalls, boots, and hairnets. As the last step of the elaborate cleaning ritual before we entered the inner sanctum, we passed through an air shower to clear off any remaining dirt or dust particles so that we would preserve the integrity of the sensitive experiments housed at SNOLAB. The entire laboratory is operated as a clean room, with the air filtered continuously; everyone and everything that enters it must be thoroughly cleaned to remove any traces of radioactive elements, which are plentiful in the mine dust and would otherwise interfere with measuring neutrino signals.

Once inside, walking by the racks of flickering electronics or having a snack in the lunchroom with a couple of scientists, it was easy to forget that there was more than a mile of rock above your head. Even if you felt claustrophobic in the elevator cage or the tunnel, you probably wouldn't here. But you might notice that there are no windows to let in sunlight. So it's perhaps ironic that this laboratory was built in the first place to peer at the Sun. Sixteen scientists came together in the mid-1980s to propose the construction of SNO to catch a handful of the neutrinos that stream out of the Sun and pass through rock more easily than sunlight through a windowpane.

Art McDonald, then a professor at Princeton University, was among them. Growing up near the eastern edge of Cape Breton Island in Nova Scotia, McDonald was always interested in how things worked. As a kid, he enjoyed taking clocks apart and trying to put them back together. Later, as a physicist, he took pleasure in applying mathematics to understand how nature worked. He returned to Canada in 1989, to take up a professorship at Queen's University and to lead the SNO project. Two years later, he and his colleagues secured sufficient funding to turn their dreams of a powerful underground neutrino observatory into reality.

The centerpiece of the SNO neutrino detector was a giant spherical vessel made of transparent acrylic. Instead of ordinary water, researchers filled it with a thousand tons of heavy water, in which deuterium atoms containing a proton and a neutron replaced hydrogen atoms with a lone proton. They purified the heavy water to remove not only dust but also any vestiges of radioactive gases. A geodesic sphere with 9,600 light sensors mounted on its inside walls surrounded the acrylic vessel, keeping a constant vigil for neutrino interactions. The whole apparatus was buried in a cathedral-size cavity deep inside the mine. When I visited the site, I could peek at it from a platform above. Building the SNO took more than nine years and over $70 million in Canadian dollars, not counting the $200 million value of the heavy water, which Atomic Energy of Canada Limited lent to the experiment. There were several snags

View of the SNO detector during installation (Lawrence Berkeley National Laboratory)

along the way, but SNO began taking data in the summer of 1999.

Two years later, Art McDonald announced the first results of their experiment after it had recorded interactions between neutrinos and the heavy water for 241 days. Comparing the number of neutrinos detected at SNO and at Super-K, his team confirmed that some must have changed their flavor. "We've solved a thirty-year-old puzzle of the missing neutrinos of the Sun," he told the media at the time. "We now have high confidence that the discrepancy is not caused by problems with the models of the Sun but by changes in the neutrinos themselves as they travel from the core of the Sun to the Earth," he said. Their results bolstered the case for neutrino oscillations and for neutrinos having at least a smidgen of mass.

This was a significant step, to be sure, but it didn't quite close the book on the problem. The cleanest test would be for SNO itself to measure all three flavors of neutrinos, without having to combine and compare with the measurements from Super-K—and that's just what the researchers set out to do next. Among other upgrades, they added two tons of sodium chloride (otherwise known as pure salt) to the heavy water. They knew that the chlorine in the salt would improve the chances of capturing neutrinos and distinguishing between the different varieties. Their clever trick paid off. Already in 2002 the team announced that the interim SNO results alone confirmed that solar neutrinos change from one type to another during their journey. The following year they reported definitive results on the neutrino numbers. The total matched what John Bahcall's solar model had predicted. Sure enough, only a third of the solar neutrinos arriving on Earth were of the electron variety. The other two-thirds were of the muon and tau types. Here was proof that electron neutrinos produced in the Sun morphed into other flavors midflight. As the Boston University physicist Ed Kearns explained, "Super-K told us just the bank balance, but SNO could actually see the record of deposits and withdrawals." The findings vindicated both Davis and Bahcall. Davis's measurements and Bahcall's calculations of solar neutrinos had been correct all along. In fact, the agreement between Bahcall's predictions and the numbers captured at SNO was surprisingly good. He was so thrilled about being proved right that he announced to a journalist that he felt like dancing. As Bahcall explained later, "For three decades people had been pointing at this guy and saying this is the guy who wrongly calculated the flux of neutrinos from the Sun, and suddenly that wasn't so. It was like a person who had been sentenced for some heinous crime, and then a DNA test is made and it's found that he isn't guilty. That's exactly the way I felt." Astronomers could finally claim to truly understand how the Sun generates power. Plus, they had a new probe of the temperature at the Sun's heart from millions of miles away, because the number of solar neutrinos produced each second is extremely sensitive to the core's temperature. Perhaps most important to the physicists, Pontecorvo had been proven correct in his prediction that neutrinos oscillate between flavors and have nonzero masses, contrary to the assumption in the standard model.

Meanwhile, perhaps inspired by these discoveries, the Nobel committee honored the momentous achievements of two pioneers of neutrino hunting. They awarded half of the 2002 physics prize jointly to Ray Davis and Masatoshi Koshiba of Kamiokande "for pioneering contributions to astrophysics, in particular for the detection of cosmic neutrinos." John Bahcall was left out, despite the dramatic confirmation of his solar model predictions. Many of his colleagues felt that was unfortunate.

Many physicists think the Nobel committee has left the door open for a future prize recognizing the discovery of neutrino oscillations. After all, the 2002 prize honored contributions to astrophysics through the detection of neutrinos, not the determination of their chameleonlike nature. Ed Kearns, for one, thinks it is "just a matter of time." Guessing the likely names of the recipients is a favorite parlor game among physicists. Kearns and several other physicists agree that the SNO team leader, Art McDonald, should get a portion. It is less clear who should receive Super-K's share of the prize, because the lead scientist, Yoji Totsuka, died in 2008. The next in line appear to be Yoichiro Suzuki and Takaaki Kajita, both of whom also played major roles in Super-K. "It would be great if the prize is given to all three. Every October I wake up wondering if it has happened," Kearns told me. John Learned of the University of Hawaii agrees with Kearns on two of the three contenders, McDonald and Kajita, but says that the third share should go to Atsuto Suzuki of the KamLAND experiment, which used neutrinos from reactors to confirm oscillations.

Several profound consequences ensued from the discovery of neutrino oscillations. For one, it showed that neutrinos were not massless, contrary to the expectations of the standard model. Thus it constituted the first bit of definitive evidence that the standard model may not be the whole story. For another, measuring those oscillations offered a way to explore "new physics," a term physicists use to describe phenomena that aren't accounted for by the standard model. As Karsten Heeger, a physicist at the University of Wisconsin–Madison, told me, "Traditional particle physics only confirmed the standard model. Neutrino oscillations were the first sign that there is something beyond the standard model. That discovery gave a huge boost to the field."

The discovery that neutrinos have mass is also of interest to cosmologists. Since neutrinos are the second most numerous particles in the universe after photons, even if each one has only a smidgen of mass, the total could add up to a lot. So some cosmologists had hoped that neutrinos would account for much of the mysterious dark matter, whose presence is only "seen" through its gravitational influence on galaxies and galaxy clusters. But the neutrino's mass has turned out to be way too tiny to explain dark matter. That means some other particle or particles, hitherto unknown to physics, must exist. The hunt is on, but no good candidate has turned up yet.

The findings of Super-K and SNO also set the stage for other neutrino experiments, focused on making precise measurements of how different neutrino flavors morph into one another. Physicists like to characterize these oscillations in terms of parameters called "mixing angles," somewhat akin to how you might describe the dynamics of an airplane in terms of pitch, yaw, and roll. They were able to determine two of these angles with Super-K and SNO, but not the third, called θ_{13} ("theta-one-three"). Measuring all three angles would allow researchers to nail down the mass differences among the three neutrino types. What's more, they may be able to glean interesting new physics from the fine details of the neutrino mutations. One thing the physicists are keen to find out is whether neutrinos and their antimatter twins behave the same way. If not, that could be the key to understanding how the universe came to be dominated by matter over antimatter, as I will discuss in chapter 7. A second question nagging at physicists is whether there really are only three flavors of neutrinos. Some exotic theories have proposed a fourth variety, called the sterile neutrino, which would never interact with matter but could be revealed by indirect means. It would be the most aloof of neutrino types, extremely difficult to pin down, yet it could be important for cosmology if it's massive enough to account for dark matter.

Most recent oscillation experiments rely on neutrinos from artificial sources, such as nuclear reactors and particle accelerators, rather than on those originating in the Earth's upper atmosphere or in the Sun, as Super-K and SNO did. Soon after the discovery of oscillating solar neutrinos, Japanese physicists confirmed the phenomenon was real by measuring antineutrinos from commercial reactors that surround Kamioka. Sure enough, only a fraction of them arrived at the detector, as if the others had morphed into different neutrino varieties along the way. Neutrino beams produced in accelerators offer experimentalists the best opportunity to control the numbers, types, and energies of the particles they study. The MINOS experiment in the United States, which shot a neutrino beam from Fermilab near Chicago to a detector in the abandoned Soudan iron mine in northern Minnesota some 450 miles away, also found evidence of neutrino oscillations.

One of the biggest neutrino oscillation experiments under way is known as T2K, for "Tokai to Kamioka." It involves sending an intense beam of neutrinos across Japan's island of Honshu. The particle accelerator where the neutrinos originate is located in Tokai on the east coast, while the detector is in Kamioka in the western part of the island, some 180 miles away. (Tokai is famous in Japan as the site of monster attacks in several Godzilla movies.) Built and operated by a large international collaboration of nearly 500 scientists from twelve countries, the experiment began taking data in January 2010. The first results were supposed to be announced at a seminar in Tokyo at 3:00 p.m. local time on March 11, 2011, but that was not to be. Just fourteen minutes before the scheduled announcement, a massive earthquake hit the northeast coast of Japan. Registering at magnitude 9 on the Richter scale, the strongest ever recorded in the country, it set off a devastating tsunami. Later estimates put the combined death toll from the quake and the tsunami at over 15,000 and the total economic cost at over $200 billion. Most worrying at the time was the meltdown at the Fukushima nuclear plant, where tsunami waves had disabled the power supply for cooling and even destroyed the backup diesel generators.

Brian Kirby, a graduate student at the University of British Columbia (UBC) in Canada, had arrived at Tokai the day before for a two-week shift in the control room. When the building began to tremble all of a sudden on the afternoon of March 11, he crouched under a table with a few others. "The shaking continued for quite some time," he recalled. Soon the power went down. Once the trembling stopped, Kirby and his colleagues stepped outside. "There were aftershocks for several more minutes, and the ground felt shaky," he said. He had no idea how far Tokai was from the epicenter of the quake, and did not realize how much devastation it had caused in and around the city of Sendai, just 125 miles to the north. Once things calmed down a bit, he and his colleagues biked to the house they had rented nearby, and had a barbecue before the food in the refrigerator went bad without power.

Hours later, back in Vancouver, Scott Oser's wife woke him up to tell him about the earthquake in Japan. Oser, a professor at UBC, is the Canadian spokesperson for the T2K collaboration, and Kirby's PhD adviser. Oser checked online maps and saw that the epicenter was not far from Tokai. Next, he opened his e-mail, hoping for a message from his student. There were many messages from T2K collaborators outside Japan, and the last one in Oser's in-box was from Kirby's worried mother. He sent her Kirby's cell phone number in Japan, and tried calling it himself. To his surprise and relief, Kirby answered. Except for the lack of power, no Internet access, and the low battery on his phone, Kirby was doing fine. The T2K collaboration arranged to evacuate their colleagues in Tokai farther inland, and later, out of the country. Kirby left Japan a few days later.

Nobody knew the condition of the experiment. "We had no expectation that anything would survive a magnitude 9 quake," Oser told me. "The roads were impassable and there was no electricity for weeks, so it took a while before we could send someone to check on the lab in Tokai." Fortunately, the damage was a lot less than feared. The buildings, which were anchored firmly to the bedrock, remained intact for the most part, but the surrounding roads had caved in, several power cables had broken, and the water supply for cooling the facility had been disrupted. Thanks to tsunami barriers, the sea waves had not reached the laboratory. Still, repairs took more than a year. The T2K experiment began taking data again in April 2012.

Meanwhile, the data collected at T2K before the earthquake, and announced in June 2011, indicated that some muon neutrinos morph into electron neutrinos. Previous research at SNO and Super-K had demonstrated two other kinds of neutrino oscillations, but this was the first time scientists saw direct evidence of the third type of transformation. Of the many muon neutrinos produced at Tokai, eighty-eight were recorded at the

Kamioka detector nearly 185 miles away. Six of the eighty-eight arrived as electron neutrinos, even though the original beam of particles consisted only of the muon variety, so they must have changed flavor midflight. As Ed Kearns explained, "Even though we have studied neutrino oscillations for years, there is still a great thrill in seeing these six events." MIT's Lindley Winslow agreed that these detections marked an exciting milestone for neutrino physics, and described them as "the six most popular neutrino events ever." With so few oscillations recorded, the T2K results were not precise enough to measure the actual value of the third mixing angle, but they showed that θ_{13} was not zero. That result opens the possibility that indeed neutrinos and antineutrinos behave differently in the way they interact with matter.

Three other experiments chasing θ_{13} were also hot on the heels of T2K. One is located in the village of Chooz in northeastern France, and measures the neutrinos produced by a commercial nuclear power plant during the course of its normal operations. Physicists have set up one detector right next to the nuclear reactor and another detector one kilometer (0.62 mile) away from it to measure the rate of disappearance of electron neutrinos. They reported the results from the first hundred days of data from Double Chooz, as the experiment is called because it involves two detectors, in the fall of 2011; their measurements provided independent confirmation that θ_{13} is not zero, but could not nail down its value very well. Another experiment, located in Daya Bay, China, is more sensitive to the transformation. Being at one of the world's most powerful nuclear plants helps, and so does having large detectors buried deep underground, to reduce the number of cosmic ray hits. After analyzing just two months of data, the Daya Bay collaboration announced in March 2012 that they had measured θ_{13} well for the first time. The researchers found that roughly 6 percent of electron neutrinos disappear as they travel the mile and a quarter between the reactors and the detectors. About a month later, a third experiment, called RENO (for the Reactor Experiment for Neutrino Oscillations), based in South Korea, confirmed the Daya Bay results, albeit with lower precision. As Kam-Biu Luk, the United States spokesperson for the Daya Bay collaboration, explains, "θ_{13} turns out to be quite sizable after all. This is a pleasant surprise." The finding, he says, "opens the floodgates for many things. It would allow theorists to extend beyond the standard model of physics."

Janet Conrad of MIT, who works on the Double Chooz experiment, is also excited about the advent of what she calls "precision neutrino physics." Growing up in northern Ohio, Conrad watched *Star Trek* as a kid and gazed at the stars through a friend's telescope. She dreamed of becoming an astronomer or a science officer on a starship. Once she started reading Nancy Drew and Sherlock Holmes mysteries, however, her career plans changed: she wanted to be a detective. Then a heavenly sight drew her back to science. As a teenager, Conrad woke up in the wee hours before dawn to spray warm water on the prize dahlias that she was growing with her father, an agricultural scientist, to display at competitive exhibitions. Out in the garden, with frigid autumn air brushing against her face, she saw the northern lights and was mesmerized by the colorful spectacle of charged particles from the Sun hitting the Earth's atmosphere. She remembers that the auroras were "so incredibly beautiful, so action-packed."

Later, as an undergraduate student at Swarthmore College, Conrad took a course in quantum mechanics and was hooked by the goings-on in the subatomic world. She got a summer job at Harvard, working in a lab that used a particle beam from a cyclotron to treat eye cancer, but it was visiting Fermilab her junior year that sealed the deal for Conrad. Today, as a neutrino hunter and a professor at MIT, Conrad combines her love of science with her fondness for solving mysteries. As she put it, "A detective is not always a scientist, but a scientist is always a detective." She believes that neutrinos hold the key to some of the biggest cosmic riddles, and is optimistic about the prospects for major breakthroughs during the coming decade. "We are entering an exciting phase. It's taken a long time to get to this point, but now we can address important questions with high-precision measurements," she adds.

Indeed, neutrino hunters have come a long way since Bruno Pontecorvo first suggested that neutrinos might suffer from a multiple personality disorder. Thanks to the discoveries at Kamiokande and SNO, these determined researchers have solved the thorny problem of the missing solar neutrinos that Ray Davis and John Bahcall wrestled with, putting their reputations on the line, for decades. Neutrino hunters have also found that neutrinos do have mass—the first definitive evidence of physics beyond the standard model—and that they morph between three flavors. Using a wide array of precision experiments, researchers have now begun to pin down the bizarre properties of these chameleon particles. In the process, they are not only pushing the frontiers of fundamental physics but also providing valuable tools for cosmology and astrophysics. In fact, next we will turn to how astronomers use neutrinos to probe the biggest explosions in the universe.

SIX

EXPLODING STARS

La Serena, about 300 miles from the Chilean capital of Santiago, is a quiet seaside town except during the peak summer months, when it attracts hordes of vacationers. Most come for the golden-sand beaches, but some also take time to enjoy the town's neocolonial architecture and the nearby countryside, renowned for vineyards making pisco, a grape brandy that is at the heart of bitter arguments between Chileans and Peruvians as to who distilled it first. North of La Serena, the fabled Pan-American Highway begins to climb, winding its way parallel to the Andes, crossing dry river valleys peppered with boulders. There is little greenery besides prickly shrubs and cactus plants. A rabbitlike viscacha might scurry among the rocks by the roadside, while a hawk might hover overhead. After nearly eighty miles, there is a turnoff to the right toward the mountains. Already visible from the junction, a set of enchanting white domes pops up against the clear blue sky in the distance. The final approach to the 7,800-foot Las Campanas summit takes would-be stargazers along a steep, narrow road carved into the hillside. The view from the top is spectacular during the day, but is absolutely stunning at night when the star-studded Milky Way arches across the sky, accompanied by two fuzzy patches of light, the Large and Small Magellanic Clouds, off to one side.

Ian Shelton, a thirty-year-old Canadian who grew up in Winnipeg, happened to be here at Las Campanas on the night of February 23, 1987. He was employed as the resident observer at a modest 24-inch telescope that belonged to the University of Toronto. In addition to taking the data that Toronto astronomers asked for, he found time to tinker with an even smaller 10-inch telescope on the mountain. Built over a half century ago and housed in a small shed, it lacked an autoguider to track the stars, so Shelton had to adjust its pointing by eye. As he often did, that night Shelton trained the little telescope on the bigger of the Milky Way's two sidekicks, the Large Magellanic Cloud (LMC), and registered long exposures of this dwarf galaxy onto old-fashioned photographic plates in order to look for variable stars in its midst.

In the wee hours of February 24, before heading to bed, he decided to develop the final plate of the night. He lifted the plate out of the developing tank and examined it, to make sure the three-hour-long exposure had come out well. Then something caught his attention: a curious bright spot next to a familiar spider-shaped feature known as the Tarantula Nebula. He wondered what the unusual spot might be, and reasoned that it was likely a flaw in the plate itself. But just to be sure, he walked out of the telescope enclosure into the dry mountain air to look up at the sky with his own eyes. He saw a bright star in the LMC that had not been visible the night before. Shelton hurried over to one of the other telescope domes on the ridge to share the news. As he discussed his puzzling find with astronomers Barry Madore and William Kunkel in the control room, the Chilean telescope operator Oscar Duhalde piped up that he had seen the same star a few hours earlier, when he stepped out for a break. Together the four of them decided the "new" star had to be a supernova, an exploding star that could briefly outshine a billion suns. No other type of astronomical object was known to change in brightness so swiftly and dramatically, from being too faint to register in photographs taken the night before to being easily spotted with the naked eye now. That meant Shelton and Duhalde had discovered a supernova in a satellite galaxy of the Milky Way.

Meanwhile, on the other side of the Pacific, in the coastal town of Nelson, New Zealand, the retired shopkeeper and avid stargazer Albert Jones was monitoring variable stars in the LMC. As a committed amateur astronomer, he took to observing the heavens with his backyard telescope whenever he could. That night he noticed a bright blue star that did not belong in the LMC, one he had never seen there before, and noted its position relative to other stars in his field of view. Before he could get a reliable estimate of how bright it was, clouds rolled in and obstructed his view. He suspected that it was likely to be a supernova, so he alerted other stargazers to his find and asked them to make follow-up observations. As it turned out, Jones had spotted the same supernova independently, mere hours after Shelton and Duhalde.

Over in Australia, at the Siding Spring Observatory about 250 miles northwest of Sydney, Robert McNaught happened to photograph the LMC on the same night as Shelton did in Chile. He promptly developed the plates with the images of the LMC, but was too busy with other chores to inspect them. When news of the supernova reached him by telephone, McNaught rushed to look at his plates. His photographs, some taken nearly fifteen hours earlier than Shelton's, also revealed the unmistakable celestial beacon. McNaught compared its position with previous images of the region, and realized that the supernova coincided with a blue supergiant star called Sanduleak −69° 202, which astronomers had cataloged and studied previously. His identification meant that, for the first time in history, astronomers knew exactly which star had blown up: in other words, they had "before and after" photographs of the exploding star, which would allow them to trace the final stages of a massive star's evolution better than ever before.

By midmorning of February 24, scientists around the world learned about the discovery, tipped off by phone calls from giddy colleagues and a telegram from the International Astronomical Union. Their delight had to do with the fact that Supernova 1987A (as it came to be known) was the first one observed in our galactic neighborhood since the invention of the telescope nearly four centuries earlier. It was the nearest and the brightest supernova seen in 383 years—since Johannes Kepler observed a supernova in our own galaxy with his naked eye in 1604—and as such, it offered astronomers an unprecedented opportunity to witness a massive star's last hurrah. Over the next hours and days, the star's debris would expand outward from the site of the initial explosion, colliding with dust and gas in the space around it. Meanwhile, the supernova would appear dimmer and dimmer in our sky.

Astronomers rushed to employ a mighty suite of optical, infrared, and radio telescopes spread across the Southern Hemisphere—particularly in Chile, Australia, and South Africa, where many advanced Earth-based observatories are located—as well as X-ray and ultraviolet instruments aboard spacecraft, all to watch the momentous event unfolding in the LMC. It was a period of frenzied activity that few scientists had ever experienced. As one ebullient astrophysicist declared, "It's like Christmas."

John Bahcall, who was an expert on models of stellar interiors and was now at the Institute for Advanced Study in Princeton, New Jersey, found it so exciting that he was losing sleep. There was a good reason for his excitement: Bahcall knew that the very first, and arguably the most important, harbingers of this cosmic cataclysm must have arrived hours *before* the astronomers using conventional telescopes spotted the supernova. He was well aware that according to theoretical models of stellar evolution, the core collapse at the end of a massive star's life should result in a copious burst of neutrinos, which flee the detonation site deep inside the star with little impediment. The visible fireworks would appear only later, when the star's outer mantle blows up. Minutes after he heard about Supernova 1987A, Bahcall and two of his colleagues got to work to calculate how many neutrinos should have been recorded by the various neutrino detectors on Earth. Initially, there was the possibility that Supernova 1987A was of the "wrong" sort—that is, not the ultimate death of a massive star, but a so-called Type Ia, which results from the explosion of a stellar cinder known as a white dwarf that has gobbled up enough material from a binary companion to reach a critical mass. But astronomical observations suggested otherwise, and the three theorists forged ahead. They determined that the "right" sort of supernova explosion should deliver a bonanza of a few dozen neutrino detections, and they submitted a paper with their conclusion to the journal *Nature* within a week so that their prediction could appear ahead of the actual measurement.

Meanwhile, experimental physicists had begun to search through data recorded at several underground detectors around the world. Their best chance of registering supernova neutrinos

Inside the Kamiokande neutrino detector (Institute for Cosmic Ray Research, University of Tokyo)

was with Kamiokande, which consisted of a huge tank of purified water, surrounded by a thousand phototubes to register
flashes of light produced when neutrinos interact with water
atoms. Fortuitously, that detector was back up and running after
a major upgrade just two months before. Many astrophysicists
were nervous with anticipation as Kamiokande team members
commenced scanning their data tapes in Tokyo. Since calculations by Bahcall's group and others suggested the apparatus
should be sensitive enough to record neutrinos from Supernova
1987A, a failure to measure them would imply a basic flaw in
our understanding of how a supernova works.

Then the results came in. To the utter relief of scientists the
world over, the neutrino signal stood out clearly in the data,
leaving no doubt as to its provenance. It meant that astrophysicists like Bahcall and his colleagues were right about what happens during a supernova explosion. The phototubes at the
Kamiokande detector had picked up eleven flashes in a burst
lasting several seconds, nearly three hours before the first optical sighting of the supernova by astronomers in Chile and New
Zealand. Halfway around the world, not far from Cleveland, a
similar neutrino detector located in a shallow salt mine under
Lake Erie registered eight flashes at exactly the same time as
Kamiokande. Later, scientists learned that a third, oil-based detector located in the Baksan Gorge of the Caucasus Mountains
in Russia had also recorded five. The two dozen neutrinos detected by these three experiments were just a few of the billions
upon billions of neutrinos sweeping past and through our planet
in a burst that originated in the heart of the exploding star in
the LMC. Since all three of these neutrino "observatories" are
located in the Northern Hemisphere, while the LMC is in the
southern sky, the neutrinos had to traverse from one side of

the Earth to the other, through our planet's interior, and enter the detectors from below. Delighted by the experimental confirmation of theoretical predictions, Bahcall told *Time* magazine that it felt surreal to be part of the scientific commotion surrounding Supernova 1987A.

Detecting a grand total of two dozen particles may not sound like much to crow about. But the significance of these two dozen neutrino events is underlined by the fact that they have been the subject of hundreds of scientific papers over the years. Supernova 1987A was the first time that we had observed neutrinos coming from an astronomical source other than the Sun. So it wasn't a big surprise that Japanese physicist Masatoshi Koshiba, leader of Kamiokande, won a share of the 2002 Nobel Prize in physics in large part for measuring these neutrinos. The ghostly particles conjured up by Wolfgang Pauli decades earlier to explain beta decay had now become useful cosmic messengers for astronomers in their quest to understand the life cycle of our Sun, as well as those of massive stars.

With these neutrino detections, as Adam Burrows of Princeton University wrote, "for the first time, we have penetrated the otherwise opaque supernova ejectum and glimpsed at the violent convulsions that attend stellar collapse." The observations confirmed the basic picture theorists had developed over decades of what happens when a massive star runs out of nuclear fuel. As John Beacom, a theoretical physicist at Ohio State University who is interested in the connections between particle physics, astrophysics, and cosmology, reflected, "Neutrinos allow us to see the interior of a massive star at the end of its life, so we can do astrophysics that astronomers could otherwise never do."

Alex Friedland of the Los Alamos National Laboratory explained that a supernova is in essence a "neutrino bomb," since the explosion releases a truly staggering number—some 10^{58}, or ten billion trillion trillion trillion trillion—of these particles. Even as astronomical numbers go, that is an astoundingly big one. In fact, the energy emitted in the form of neutrinos within a few seconds is several hundred times what the Sun emits in the form of photons over its entire lifetime of nearly 10 billion years. What's more, during the supernova explosion, 99 percent of the precursor star's gravitational binding energy goes into neutrinos of all flavors, while barely half a percent appears as visible light.

Since the Large Magellanic Cloud is located some 160,000 light-years away, the neutrinos generated in the supernova that reached the Earth in 1987 began their journey 160,000 years ago, at a time when early *Homo sapiens* roamed East Africa and woolly mammoths tramped the Siberian tundra. The fiery saga of the star Sanduleak −69° 202 itself began roughly 11 million years earlier, around the time when grassland mammals started to spread across the Earth but before the rise of the Himalayas. For the first 10 million years of this star's life, it was fueled by nuclear reactions that convert hydrogen into helium, just like our Sun. The energy released by these reactions prevented the star from collapsing inward under its own weight. When its core had turned almost entirely into helium, there was a pause in energy production. No longer able to fight off gravity, the core shrank and heated up, while the star's outer layers expanded as hydrogen continued to burn in a shell around the core. Once the density and the temperature of the core were high enough, the helium began to fuse into carbon and oxygen. By then, Sanduleak −69° 202, originally about twenty times more massive than the Sun, had become a red supergiant, swelling to five hundred times the Sun's size.

The star continued to burn helium in its core for nearly a million years, until that supply was also exhausted and gravity took over again. Next, according to what we know from stellar evolution models, the core shrank until it was sufficiently dense and hot to fuse carbon into neon, sodium, and magnesium. Around this time, we think the star likely expelled some of its distended outer envelope, while the remaining body shrank a bit, changing its color from red to blue. Now the pace of its evolution quickened even more. Carbon fusion went on for about 12,000 years. Next the star burned off neon and oxygen, each phase lasting for just a few years. Finally, silicon and sulfur in the core combined to make iron in about a week. By then, we expect, the star resembled a gigantic onion, with layers of different elements wrapped around an iron core. That was the end of the road, because iron cannot fuse into heavier elements without absorbing energy from the outside. The star could no longer resist the crush of gravity. When the end came, it was quick and spectacular, with the bonfire visible to the unaided eyes of humans living on a small rocky planet in the outskirts of the neighboring Milky Way galaxy, 160,000 years after the calamity.

Scientists think neutrinos had a lot to do with hastening the demise of Sanduleak −69° 202. The radiation emitted during the carbon-burning stage of the star, when the core temperature reached 500 million degrees, was so energetic that it would give rise to electron-positron pairs (since energy can transform into matter and vice versa, à la $E = mc^2$, as Einstein suggested). Typically, these particle-antiparticle pairs would annihilate each other when they met, producing gamma rays, but sometimes they would create neutrino-antineutrino pairs instead. Since neutrinos and antineutrinos rarely interact with their surroundings, they would have escaped from the star, draining it of energy that would otherwise help resist the inward crush of gravity.

And there's more: neutrinos may play a starring role during the supernova event itself. Once the spent star's iron core reaches a critical mass of 1.4 times that of the Sun (a threshold known as the Chandrasekhar limit after the Indian astrophysicist who came up with the concept), it collapses down to a sphere only thirty miles across within a fraction of a second. The extreme temperatures mean that there is plenty of energy to create more neutrino-antineutrino pairs. These particles sneak out, taking several seconds to do so because of the incredible density of the core, and carry away lots of energy with them. Meanwhile, free neutrons, which are plentiful in this energetic environment, combine with iron nuclei and forge even heavier elements. The implosion stops when this inner core gets as dense as an atomic nucleus, when the nuclear force prevents protons and neutrons from cramming any closer together. In fact, the shrinking core bounces back, colliding with the infalling outer layers and generating a powerful shock wave. But, according to state-of-the-art computer simulations, the shock wave runs out of steam as it travels outward.

That's when neutrinos might step in again and lend a helping hand. "If a small fraction of neutrinos streaming out of the core dump their energy in material just behind the stalled shock, that's enough to get the shock wave moving again," explains Georg Raffelt of the Max Planck Institute for Physics in Munich, Germany. "If not for neutrinos," he stresses, "the whole thing would collapse into a black hole, perhaps without visible fireworks." The revived shock blows apart the star's outer layers.

As a result, heavy elements up to and including iron, produced during the star's lifetime, as well as even heavier elements fused during the supernova event, are expelled into space. Some of the enriched material gets incorporated into new generations of stars, their planetary systems, and eventually our bodies. The calcium in our bones, the iron in our blood, and the oxygen we breathe all came from ancient supernovae. So did the copper we use to make wires; the silver, gold, and platinum that we use for jewelry; and the gallium that we use in electronic components. As Raffelt points out, "Since neutrinos are crucial for blowing up stars, they are important for our very existence." If not for neutrinos, the universe might well have been a barren wasteland, devoid of much of its splendor, and unable to give rise to life.

The handful of neutrinos detected from Supernova 1987A, combined with astronomical observations, provided broad support for the scenario that theorists had developed, with the help of complex simulations on supercomputers, for how an aging massive star self-destructs, with its core collapsing into a tightly packed ball of neutrons—called a neutron star—or a black hole, and its expelled outer layers forming a glowing cloud of debris. When we look at images of the site taken with the Hubble Space Telescope today, we can see a bright ring as well as two loops that appear to interlock, presumably material ejected by the progenitor star and lit up later by ultraviolet light from the supernova. But there is a crucial piece missing in this picture. Given the estimated mass of the progenitor star, astronomers expect its core should have formed a neutron star, but they haven't been able to identify it. They think the corpse might be hiding behind the dust.

The supernova neutrino detections, as sparse as they were,

also validated some important details of the story of how a massive star blows up at the end of its life. The astrophysicists involved in the research were pleased to find that the number and energies of the neutrinos they measured agreed with their expectations based on theoretical calculations of the explosion. Thanks to this excellent agreement between theory and observation, researchers concluded that the supernova didn't lose energy through some mysterious process. For example, they were able to rule out speculations about neutrinos emitting hypothesized exotic particles called axions or leaking into enigmatic extra dimensions. The arrival of neutrinos over several seconds, rather than all in a single burst, confirmed that they took some time to make their way out of the extremely dense, shrunken core, as predicted.

Besides all the insight they provided into supernova dynamics, the measurements also revealed clues to the nature of neutrinos themselves. Since the neutrinos reached Earth no more than three hours before the supernova was captured in an optical photograph, they must have traveled at a speed pretty close to that of light. Since lighter particles travel faster than heavier ones, scientists reasoned that the mass of a neutrino must be quite small. In fact, based on their arrival time from Supernova 1987A, scientists were able to show that, despite their prodigious numbers, neutrinos would be unlikely to account for the mysterious "dark matter" permeating the universe. What's more, as we saw in chapter 1, when a media frenzy broke out in 2011 about neutrinos traveling faster than light, one strong counterargument was based on these observations. If these particles indeed travel as fast as the OPERA experiment initially reported, the neutrino burst from Supernova 1987A should have reached the Earth years earlier than optical light, not a mere three hours before.

Supernova 1987A whetted the appetite of astrophysicists who want to learn about the inner workings of dying stars. "Imagine what we could learn if we were to detect a thousand neutrinos from a nearby supernova," muses Alex Friedland. Such a prodigious event would not only allow us to pinpoint the sequence of events as the explosion proceeds but also tell us definitely whether the end result is a black hole or a neutron star. Particle physicists are also interested in neutrinos from supernovae, because they provide a rare opportunity to understand how these elusive particles behave under extreme conditions that can't be replicated in a laboratory.

What both sets of scientists need to achieve their goals is a core collapse supernova in our own galaxy. However, no supernova has been seen in the Milky Way since 1604, when stargazers including the German mathematician Johannes Kepler noticed a "new star" in the constellation Ophiuchus. At its peak, this supernova was so bright that it was visible during the daytime. Modern radio, optical, and X-ray telescopes have observed the remnant it left behind in the form of a shell of hot gas. Just three decades before Kepler saw the 1604 supernova, observers in Europe had seen one. The legendary Danish astronomer Tycho Brahe, who saw the earlier supernova in 1572, wrote that upon spotting "a new and unusual star, surpassing all others in brilliance" in the constellation Cassiopeia, "I was so astonished at this sight that I was not ashamed to doubt the trustworthiness of my own eyes. But when I observed that others, too, on having the place pointed out to them, could see that there was really a star there, I had no further doubts." In fact, Tycho's meticulous observations of the 1572 supernova helped establish that the heavens were not immutable, as Aristotle had argued and many Western philosophers still believed in the sixteenth century.

Historical records and ancient rock engravings indicate that a number of other supernovae were seen with the naked eye over the last few millennia. The earliest probable account from China, seared on an animal bone dating back to 1300 B.C.E., refers to "a great new star" that was seen near the bright red star now known as Antares. The Chinese chronicle titled *Book of the Later Han* provides a description of a "guest star" that appeared in the year 185 and faded slowly over many months. European monks and an Egyptian astrologer were among those who noted the brilliant supernova of the year 1006, which may well have been the brightest stellar event ever recorded. The chronicler of a Benedictine monastery at St. Gallen in Switzerland described the sight: "A new star of unusual size appeared, glittering in aspect, and dazzling the eyes, causing alarm . . . It was seen likewise for three months in the inmost limits of the south, beyond all the constellations which are seen in the sky." The scholar Ali ibn Ridwan, based in Cairo, wrote, "This spectacle appeared in the zodiacal sign Scorpio, in opposition to the Sun . . . The sky was shining because of its light. The intensity of its light was a little more than a quarter that of moonlight." The supernova of 1054, with expanding debris that forms the famous Crab Nebula in the constellation Taurus, is mentioned in Chinese, Japanese, and Arab historical accounts, and some scholars argue that Native Americans of the Southwest also depicted it in rock engravings.

In the 1930s, Caltech astronomers Walter Baade and Fritz Zwicky recognized that supernovae had to be enormously bright to be seen across intergalactic distances. Born in Germany, Baade was a meticulous observer and a gentle, soft-spoken man. He met Wolfgang Pauli in Hamburg, and they became lifelong friends.° The two even wrote a paper together on the curved shape of comet tails. Baade moved to California in 1931 to work at the Mount Wilson Observatory. Classified as an enemy alien during the Second World War because of his German nationality, he spent many nights using the 100-inch Hooker telescope, the world's largest at the time, to obtain clear photographs of faint galaxies, taking advantage of the wartime blackouts of nearby Los Angeles. Building on the work of Edwin Hubble, Baade established that the universe was much bigger than previously thought.

Unlike his colleague, Zwicky was cranky and overbearing, and was fond of disparaging his enemies as "spherical bastards" (spherical, he said, because they looked like bastards from every angle). Born in Bulgaria to Swiss parents, Zwicky grew up with his grandparents in Switzerland, where he later met Wolfgang Pauli and Albert Einstein. After completing his PhD in Zurich, he went to Caltech on a fellowship and stayed on as a professor. He was an avid skier and mountain climber, and competitive both in sports and science. Among his wide-ranging discoveries was the finding that "dark matter" dominated the mass of galaxy clusters. He also proposed that nearby galaxies could act as "gravitational lenses," bending and amplifying the light of more distant ones along the same line of sight. Somehow Baade and Zwicky managed to collaborate, at least for a while, despite their very different personalities (though later Baade feared that Zwicky might try to cause him physical harm).

In a prescient paper published in 1934, Baade and Zwicky

° It was to Baade that Pauli made his famous confession about invoking the neutrino: "I have done a terrible thing. I have postulated a particle that cannot be detected."

wrote, "With all reserve we advance the view that a super-nova represents the transition of an ordinary star into a *neutron star*, consisting mainly of neutrons. Such a star may possess a very small radius and an extremely high density." Their insight seems all the more remarkable when we recall that physicists had discovered the neutron barely two years earlier. Next, Zwicky set out to find lots of supernovae by surveying a large number of galaxies with a wide-field telescope. Over his lifetime, he identified more than 120 supernovae.

Based on their observations of other galaxies, today's astronomers estimate that at least a few massive stars must explode per century in the Milky Way. But we could miss these supernovae easily if they occur far from our locale, because the dust in interstellar space obscures our view of the distant realms of our galaxy. In fact, recent radio and X-ray observations have revealed the remnant of an explosion toward the center of the galaxy, which should have been seen as a supernova about 150 years ago but went undetected. Even if interstellar material blocks light from a supernova, it does not hinder the passage of neutrinos, so detecting a burst of neutrinos would reveal the death of a massive star anywhere in the Milky Way. We have had sensitive neutrino detectors operating for about a quarter century, but they haven't registered a galactic supernova yet. "It would be a once-in-a-lifetime opportunity, so we better be prepared," Raffelt advises.

Kate Scholberg of Duke University agrees. She and her colleagues have set up the SuperNova Early Warning System, or SNEWS for short, a coordinated network to provide rapid notifications of core collapse explosions in the galaxy. The plan is that detectors around the globe that are sensitive to supernova neutrinos (such as IceCube in Antarctica, the Large Volume

Detector and Borexino in Italy, and Super-Kamiokande in Japan) will report candidate bursts to a central computer at the Brookhaven National Laboratory on Long Island, New York. "If several neutrino detectors light up at once, there's a very good chance a supernova has gone off nearby," Scholberg explains.

If the SNEWS computer finds a coincidence within ten seconds between signals from two detectors, it sends out an alert to observatories worldwide. For maximum speed, the alerts go out without human intervention. Scholberg and her colleagues hope that telescopes on the ground and in space will be able to record electromagnetic radiation, including visible light, radio waves, and X-rays, from the explosion sooner rather than later, and watch its early stages unfold. There is just one hitch: most neutrino detectors can't tell precisely which direction the particles come from, so astronomers won't know where to look. "Still, the alert will enable telescopes with large fields of view to initiate searches. Plus, we have lots of amateur astronomers signed up—many of them have intimate knowledge of the sky," says Scholberg. "The idea is to have as many people looking as possible, everywhere, to have the best chance possible of pinpointing early light."

Scholberg points out that "measuring neutrinos from a galactic supernova will tell us an enormous amount. It's an unbelievably rich mine of information." The detectors will record how the number and energy of the arriving neutrinos evolve over time, which will give us insight into how the explosion unfolds. Among other things, scientists will be able to determine whether the star's core collapses all the way into a black hole, from which nothing—not even neutrinos—can escape, or stops short, forging a neutron star instead. If a black hole was to form, the stream of neutrinos from the supernova would come to a sudden halt. If the end product was a neutron star, on the other hand, the stellar cinder would continue emitting neutrinos for about ten seconds while it cooled down, so the neutrino stream should dwindle slowly instead of being cut off abruptly. In the latter case, as Scholberg explains, "we will get to see the initial cooling of a neutron star, and learn about properties of super-dense matter."

A galactic supernova could also shed light on the nature of neutrinos, and help us address some of the unresolved questions that I discussed in the last chapter. For example, physicists have been struggling to determine what they call the "mass hierarchy" of neutrinos. In effect, they want to know if there are two heavy-mass states plus one light state, or one heavy and two light states, and they believe that measuring supernova neutrinos would nail down the answer. What's more, in a supernova core the number density of neutrinos is so high that interactions between neutrinos, which are otherwise oblivious to each other's presence, could alter their behavior. "We might see some exotic collective oscillations of neutrinos," says Scholberg. "If there are any anomalies in their behavior, they could point to new physics beyond the standard model." John Beacom concurs: "We might learn something about neutrinos that we can't measure in the lab."

Fortunately, several existing detectors, such as Super-Kamiokande, Borexino, and IceCube, are sensitive to neutrinos from a supernova that occurs anywhere in the Milky Way. Super-Kamiokande, for example, would register several thousand events from a supernova near the galactic center, over 25,000 light-years away. It could even pinpoint the direction the neutrinos come from to within a few degrees, corresponding to a patch of the sky several times bigger than the full moon. IceCube, which would record a million events, is best for tracking how the neutrino stream evolves with time, because it can measure the flux in time steps as short as a few thousandths of a second. "We will be able to see the entire ten-second story of the explosion unfold in snapshots taken every few milliseconds," says Ice-Cube principal investigator Francis Halzen of the University of Wisconsin–Madison. "We will be able to pin down the exact moment that the neutron star forms."

The current detectors are only sensitive to one flavor of neutrinos, namely electron antineutrinos. Even though supernovae also release electron neutrinos, as well as muon neutrinos and tau neutrinos plus their antimatter counterparts, these detectors wouldn't register any of them. Scientists would like to measure all three neutrino flavors, and their antimatter doppelgängers, of course. "Observing only one flavor is like taking a photograph through a single-color filter," explains Scholberg. She would rather have the full-color view. As a first step toward developing multicolor vision, Scholberg and Canadian colleagues are building a dedicated apparatus, called the Helium and Lead Observatory (HALO), at the SNOLAB in northern Ontario, Canada. Using 80 tons of lead as the detector material, HALO is uniquely sensitive to electron neutrinos, so it will be complementary to other existing detectors that register their antimatter twins. HALO is fairly small as neutrino detectors go, so a supernova would have to explode within the nearer half of the galaxy to be detectable. With little idea of when the next galactic supernova will occur, it's difficult to justify building a larger neutrino observatory dedicated to that purpose. "Your detector has to have a day job while it awaits a supernova," Scholberg explains.

That is indeed the plan for the Long-Baseline Neutrino Experiment (LBNE). Proposed for construction in the abandoned Homestake Gold Mine in South Dakota, it would use a gigantic tank filled with 30,000 tons of frigid liquid argon to trap a beam of neutrinos or antineutrinos sent through the ground from Fermilab, 800 miles away, to measure how these particles oscillate among their three flavors. But LBNE would also be sensitive to the different types of neutrinos coming from a galactic supernova. "Measuring these different flavors, and their time evolution, will give us tremendous insight about all kinds of phenomena," says Scholberg. "We will learn not only about the conditions in the supernova core but also about the nature of neutrino oscillations."

For instance, when protons and electrons fuse together to form neutrons in the supernova core, the result is a burst of particles that consists almost entirely of electron-flavor neutrinos. But on their way out of the core, they may change, or oscillate, into other neutrino types. "So if we observe this initial sharp spike is made of different flavors, not just electron neutrinos, that tells us a lot about oscillations," Scholberg explains. Unfortunately, as I'll highlight in chapter 8, the U.S. Department of Energy has approved funding only for a smaller, barebones version of the LBNE, with limited capabilities. Meanwhile, European and Japanese physicists have proposed their own versions of neutrino observatories that will be sensitive to all three flavors of neutrinos, and will come in handy when a supernova next explodes in our galaxy.

Theorists, for their part, are fine-tuning the supernova models, with the help of complex computer simulations. "Next-generation supercomputers may be needed to track the problem of what happens in the first second of the explosion," says Alex Friedland. But, he explains, "I think the current calculations are reliable as to what happens over the next few seconds." Predicting how the neutrinos would interact with each other and change flavors within the remarkably dense supernova core is particularly challenging. "You need to do quantum mechanics for an entire ensemble of particles," says Friedland.

Meanwhile, an entirely different sort of "observatory," one that could provide unprecedented insights, especially when combined with neutrino detections, is also getting ready for a nearby supernova. The Laser Interferometer Gravitational-Wave Observatory (LIGO) is based at two sites some 1,850 miles apart, in Hanford, Washington, and Livingston, Louisiana. At each location, there is an L-shaped tunnel system, with arms extending perpendicularly for two and a half miles. Inside each arm, a laser beam shines end to end through a long vacuum tube. Precise instruments in the central building can detect the slightest change in the distance traversed by the lasers. If LIGO was to register a path length difference as small as one-thousandth of a proton's width between the two arms, that could mark the arrival of gravitational waves, ripples in the fabric of space itself caused by a distant cataclysm. Such squishing and stretching of space was predicted by Albert Einstein's theory of gravity, but has never been seen directly. The gravitational waves are so subtle that even a passing truck produces much stronger vibrations than two neutron stars colliding in space. That's the reason LIGO has set up identical equipment at two sites far apart, to be able to discern a genuine cosmic signal from the many local sources of noise.

When I visited the Hanford site in 2011, after driving from Seattle across the Cascade Mountains on a snowy day, the observatory was in the midst of a major upgrade. Once the upgrade is completed in 2014, Advanced LIGO, as it is known, should be sensitive to the gravitational disturbance caused by two neutron stars colliding as far as *a billion* light-years away. It should also "hear" the death shriek of a massive star in our cosmic neighborhood—*if* the explosion is off balance. If the collapse proceeds smoothly and symmetrically, not a whimper will be heard, because gravity waves emitted symmetrically cancel each other out. But if the collapse were chaotic, proceeding faster in some directions than others, it would produce a strong gravitational wave signal. That could well be the case if the shrinking core of the star is rotating furiously, so that it takes the shape of a football.

We do have evidence that explosions can be asymmetric: astronomers have observed some neutron stars, presumably formed in supernovae, racing through space at speeds as high as a few hundred miles a second. Additionally, soon after a neutron star's birth, its dense nuclear matter may slosh about, buffeted by the star's frenetic spin, producing gravitational waves as it does so. "Detecting both gravitational waves and neutrinos from the same supernova event would be extremely interesting," says Kate Scholberg. Computer simulations suggest that combined data will allow scientists to measure how fast the collapsing core rotates, and also to pin down the details of the explosion mechanism. These detections would mark the dawn of true "multi-messenger astronomy," when scientists can glean complementary information from electromagnetic radiation as well as neutrinos and gravity waves.

As exciting as the prospects are, realizing them will have to wait until a core-collapse supernova goes off in the galactic neighborhood. Both Scholberg and Beacom admit the long wait is frustrating. As Beacom says, it is "a matter of holding your breath." The problem is that current observatories are not sensitive enough to detect many neutrinos from supernovae in other galaxies. For example, Super-Kamiokande would register a single paltry event from an explosion in the Andromeda galaxy, the Milky Way's nearest big neighbor, two and a half million light-years away. With a much larger future detector, such as the proposed LBNE, the number of events would rise to a few dozen—nowhere near enough to satisfy eager neutrino hunters.

Beacom and his colleagues are charting a path less traveled: they hope to catch a glimpse of the sea of relic neutrinos left over from all the core-collapse supernovae that have occurred throughout the entire cosmic history. On average, there is a star exploding every second somewhere in the universe, so there should be plenty of supernova neutrinos swarming about. Beacom is looking for the sum total of these neutrinos, a quantity that scientists call the "diffuse supernova neutrino background." He estimates that several hundred relic neutrinos fall on each square inch of the Earth per second, a puny number compared to solar neutrinos and those produced in the Earth's atmosphere through cosmic ray interactions. According to Beacom, "It's a very weak signal, but we think we have a good chance of detecting it soon." The challenge is to distinguish the supernova relic neutrinos from their much more numerous local counterparts. Beacom and his colleagues have suggested that dissolving a bit of gadolinium, a silvery-white metal, in the the giant water tank at Super-Kamiokande would do the trick, since the fix would enhance the detector's sensitivity to relic neutrinos. Their goal is to understand what neutrino emission from a typical supernova looks like without having to wait for many fresh individual examples.

Of course, detecting the faint blended neutrino signal of many supernovae is no substitute for tracing the intricacies of a nearby explosion as it unfolds. When we consider the Sun's environs, one good candidate for a supernova progenitor is Betelgeuse, a red supergiant star that marks the hunter's right shoulder in the Orion constellation, located a mere 640 light-years from Earth. This star is so bloated that if it were to replace the Sun, it would engulf the Earth's orbit, and its outermost layers would extend beyond Mars. If Betelgeuse were to blow up, the resulting supernova could shine as brightly as the full Moon in our skies for days or even weeks. What's more, the Super-Kamiokande detector would register 60 million events over several seconds, causing a different sort of headache. As Raffelt explained, "Neutrino detectors are usually designed for low event rates, so too many hits could make them go blind because the electronics won't be able to cope."

Eta Carinae, a rare behemoth of a star at least a hundred times more massive than our Sun, is another supernova contender. Located some 7,000 light-years from us, it is a bizarre and volatile beast. It has displayed wild swings in brightness over the past three centuries, flaring up to become one of the brightest stars in the night sky in 1843—when it was mistaken for a supernova at first—and remaining so for two decades. That episode was associated with a violent outburst, when the star belched out its outer layers, losing a tenth of its mass. The ejected material now forms two gigantic lobes of gas, which look like two big balloons with the star in the middle. Eta Carinae is surely on a roller-coaster ride toward a glorious death. The next detonation may well be its last. Given how hefty it is, the corpse it leaves behind will be a black hole. If Eta Carinae were to explode as a supernova anytime soon, detectors on Earth would record about half a million neutrinos.

While all the evidence suggests Betelgeuse and Eta Carinae

will meet fiery ends in the near future, we don't know what "near" means. In cosmic terms, it could well be several hundred thousand years from now. That said, the odds are pretty good that a massive star somewhere in the galaxy will explode in the next few decades. As Alex Friedland told me, "If I had to bet on what would happen first, the next galactic supernova or building of the next big particle collider in the U.S., my money would be on the supernova." Even if the supernova was so far away from the Earth that we couldn't observe its light through the dusty veil of the Milky Way, it would shine brightly in neutrino detectors around the world. It would be a sensational event, a watershed moment in their quest that neutrino hunters will celebrate like no other.

SEVEN

VANISHING ACTS

We have all seen astronomers on television musing eloquently about the vast emptiness of the cosmos. Of course, there are galaxies, stars, planets, and people within the immensity of space, so it is not quite empty. For physicists, that obvious fact is, ironically, a veritable mystery—because the universe could well have ended up with no matter at all. They are now turning to neutrinos to understand how the universe avoided such a dismal fate.

The cosmic birth in the big bang created a vast quantity of energy. The newborn universe was compact, dense, and hot, with plenty of energy for the spontaneous creation of particle and antiparticle pairs. The cosmic density was so high that these pairs should have come together and destroyed each other quickly, leaving only a sea of radiation. Since that is not the case, physicists have concluded that there must have been a slight preponderance of matter over antimatter. They have worked out the magnitude of that asymmetry: for every billion antiparticles, there should have been a billion and one particles. All objects in the universe today—including us—owe their existence to that tiny excess of matter back then.

So how did a slight bias for matter over antimatter come about? That question is among the most basic, yet elusive, in particle physics. Its solution has been the goal of much theoretical work as well as quite a few experiments for several decades. Physicists have wondered whether they will find a solution to this deep mystery within the standard model, or whether they will need to invent a whole new theory of matter. The model has been remarkably successful at explaining the complex world around us in terms of a handful of fundamental particles—plus their antimatter doppelgängers. It describes the various interactions between these particles through the exchange of "force carriers" such as photons. Countless experiments have verified the standard model's predictions to incredible precision. But the origin of the cosmic favoritism for matter over antimatter has remained a thorn in the model's side. As Edward "Rocky" Kolb of the University of Chicago explains, "It could be that this asymmetry really points to something beyond the standard model." Indeed, Kolb believes this is why particle physicists are so interested in solving the riddle of the missing antimatter. Mounting evidence points to neutrinos, or at least their massive counterparts in the early universe, as the likely culprits in this cosmic whodunit.

To understand the physics behind the mystery, let us recall that atoms contain members of two big particle families, called baryons and leptons. Baryons, such as protons and neutrons, are made up of even smaller particles known as quarks, which come in six types, or flavors. The strong nuclear force holds the quarks together. What's more, each quark is characterized by a baryon number, a charge, and a "color." Leptons, on the other hand, are fundamental particles in themselves: they can't be broken down any further. Electrons and neutrinos belong to the lepton family. A lepton number and a charge identify each lepton.

According to the standard model, each fundamental particle has an antimatter twin, with the same mass but opposite charge and spin. For example, an electron and a positron have the same mass, but one has a charge of −1 while the other has +1 charge. Just as quarks combine to make baryons, antiquarks can come together to form antibaryons. When matter and antimatter come into contact, they annihilate each other, producing photons. The "rules" of the standard model suggest that interactions between particles conserve the baryon number and the lepton number: in other words, the total baryon and lepton number of everything you started with should equal the total baryon and lepton number of all that's left at the end. But if these rules were strictly true, we wouldn't be here! At some point in the very early universe, there had to be some reactions that didn't obey these conservation rules perfectly, in order to account for the dominance of matter over antimatter that we see today.

The story of antimatter began with Paul Dirac, whom Stephen Hawking once described as "probably the greatest British theoretical physicist since Newton." Dirac was born in 1902, in Bristol, in the southwest of England, where his father, a Swiss immigrant, was a French teacher and his mother was a librarian. Dirac didn't get along well with his father, who was rather strict and demanded that his children speak to him only in French. As a result of his father's authoritarian manner, Dirac's childhood was not a happy one. Dirac once said, "I never knew love or affection when I was a child." After completing degrees in electrical engineering and mathematics at the University of Bristol, he pursued a doctorate in physics at Cambridge University, where he later became a professor. Dirac's strange habits became the stuff of legend. The Indian-born astrophysicist Subrahmanyan Chandrasekhar, who took a course from Dirac at Cambridge, described how he would go "slyly along the streets," walking "quite close to the walls (as if like a thief!)." An avid mountaineer, Dirac was sometimes seen climbing trees in the hills near Cambridge, wearing the same black suit that he donned for lectures on campus, presumably to prepare for climbing excursions. Despite his eccentricities, Dirac was a loving family man who enjoyed riding bicycles, swimming, and canoeing with his children.

Dirac was known for speaking so sparingly that his colleagues jokingly defined "the Dirac" as a unit of measurement for the fewest words a person could utter per hour while still taking part in a conversation. Quantum physics pioneer Niels Bohr grumbled about his tremendous reticence: "This Dirac, he seems to know a lot of physics, but he never says anything." There are many anecdotes about Dirac's unusual personality, especially his excessively logical and literal-minded reactions, which some have suggested might be evidence of autism. Once during a conference held at a castle, one participant remarked that a ghost supposedly haunted one of the rooms at midnight. Dirac is reported to have asked, in all earnestness: "Is that midnight Greenwich time, or daylight saving time?" The story of Dirac and Werner Heisenberg, a founder of quantum physics best known for proposing the uncertainty principle, taking a cruise ship to attend a conference in Japan also illustrates Dirac's matter-of-fact attitude. The gregarious Heisenberg took part in the dances on board the ship, while Dirac sat quietly in a corner. "Why do you dance?" Dirac asked Heisenberg at one point. "When there are nice girls, it is a pleasure," he replied. After a long silence, Dirac asked: "But how do you know be-

forehand that the girls are nice?" Dirac was also known for his
critical views of religion, and especially for questioning its political purposes. After hearing a sharply worded critique he
made in a discussion of the religious views of physicists, Wolfgang Pauli remarked, "Well, our friend Dirac has got a religion
and its guiding principle is 'There is no God and Paul Dirac is
His prophet.'" Everyone present burst into laughter, including
Dirac.

While at Cambridge, in 1928 Dirac wrote down a mathematical equation that combined the newly developed theories
of special relativity and quantum mechanics to describe the
behavior of the electron. The equation was not only elegant in
its simplicity, but it also predicted various properties of the
electron correctly. To Dirac's own surprise and initial chagrin,
however, the equation also implied that there must be a positively charged version of the electron. At first he thought that
the proton could fit the bill. After all, electrons and protons
were the only two types of matter particles known to scientists
at the time. But his equation seemed to require exact symmetry between the two versions: the positive "mirror" particle must
have the same mass as the electron. Since the proton was nearly
two thousand times more massive than the electron, the two
didn't appear to make a good pair.

By 1930, other scientists also pushed back against Dirac's
initial suggestion that the proton and the electron might be mirror particles of each other. In particular, J. Robert Oppenheimer,
who went on to lead the Manhattan Project to develop a nuclear
bomb during the Second World War, and the Russian physicist
Igor Tamm found an even more serious problem with this interpretation. Working independently, they realized that if two
particles of the opposite types described by Dirac's equation

were to come together, they would destroy each other, producing a burst of energy, through a process physicists now refer to as annihilation. If the proton were the positive counterpart of the electron, atoms wouldn't be stable, because the two types of particles couldn't survive near one another. The only way out, as Dirac himself suggested the following year, was that his equation posited a new particle, which he dubbed the antielectron.

In fact, Dirac realized that his equation called for the existence of "an entirely new kind of matter." For every particle there had to be a corresponding antiparticle, sort of a mirror image with the same mass but opposite in other respects, such as electric charge. What's more, Dirac's equation also implied that given enough energy, particle-antiparticle pairs could pop into existence spontaneously, something that physicists found difficult to believe at first.

Just a few months later, while studying cosmic rays (the energetic particles arriving from outer space), Carl Anderson of the California Institute of Technology noticed a track left in a cloud chamber by "something positively charged, and with the same mass as an electron." After about a year of investigations, he decided these new particles were indeed antielectrons, and he called them positrons. Anderson also found instances when electron-positron pairs seemed to appear out of nowhere, confirming that pair production is a real physical process, just as Dirac's equation had predicted. There is free lunch after all, at least for a limited time before the pair self-destructs. Nowadays, scientists produce millions of electron-positron pairs routinely in accelerator experiments, and they use magnetic fields in a vacuum chamber to separate the two types before they have a chance to interact with and annihilate each other.

Dirac was awarded a Nobel Prize in 1933, at the age of

thirty-one, for his theoretical prediction, which had been proved right by Anderson's discovery of the positron. Exceedingly shy man that he was, Dirac considered turning down the prize to avoid the public attention it would bring. "As shy as a gazelle, and as modest as a Victorian maid," was how the London newspaper *Sunday Dispatch* described Dirac at the time. But his friends persuaded him that declining a Nobel would draw even more publicity, and so he attended the ceremony.

Even though Anderson identified the positron soon after Dirac's prediction, it took a little longer to identify the antiproton and the antineutron. Emilio Segrè and Owen Chamberlain spotted the antiproton in a particle accelerator named the Bevatron in Berkeley, California, in 1955. A year later, Bruce Cork and his colleagues discovered the antineutron at the same accelerator. Two independent research teams observed the first antinucleus, made of an antiproton and an antineutron, in 1965, but it wasn't until thirty years later that scientists were able to produce the first antiatoms, with a positron orbiting an antinucleus. In 1995, a team of physicists built up nine "antihydrogen" atoms at CERN. Efforts to make large quantities of antihydrogen and store them long enough to study their behavior are still continuing at CERN. By colliding gold nuclei at high speeds inside an accelerator at the Brookhaven National Laboratory, to simulate densities similar to those of the universe microseconds after the big bang, researchers were able to produce antimatter versions of helium nuclei, made of two antiprotons and two antineutrons, for the first time in 2011. These were the heaviest antimatter nuclei made yet.

Antimatter is not only difficult to make in the laboratory but also rare in nature: today's universe appears to be made almost entirely of matter. How do we infer there isn't a lot of antimatter out there? First, we can be quite sure there isn't much antimatter in the solar system. (After all, the astronauts who landed on the Moon and the unmanned probes sent to planets, asteroids, and comets didn't explode upon contact.) Particles in the solar wind don't annihilate when they reach Earth's atmosphere, which tells us the Sun is also made of matter just like the Earth. Among the energetic cosmic rays that originate from the far reaches of the Milky Way and bombard the Earth every day, protons outnumber antiprotons 10,000 to 1, which suggests that there isn't much antimatter elsewhere in our galaxy either. It is also extremely unlikely that there are other galaxies made of antimatter, because if there were, we should see strong and widespread gamma ray emission whenever they interact with their matter-dominated counterparts.

Some scientists have attempted to measure directly just how rare antimatter is in the universe. Samuel Ting, a particle physicist at MIT, and his collaborators have built an instrument called the Alpha Magnetic Spectrometer (AMS), which uses a huge superconducting magnet and six sensitive detectors to look for antihelium nuclei among cosmic rays. The prototype of Ting's instrument, which NASA flew aboard the space shuttle *Discovery* in 1998, detected millions of helium nuclei, but none of their antimatter twins. Astronauts installed the full-scale experiment, which is a thousand times more sensitive than the prototype, on the International Space Station during the shuttle *Endeavor*'s final flight in 2011.

Scientists have puzzled over the paucity of antimatter in the universe for decades, because it suggests the breaking of symmetry at a fundamental level. The concept of symmetry is central to physics, and so are laws of conservation. Back in 1915, the remarkable German mathematician Emmy Noether realized that the two concepts of symmetry and conservation are intimately linked. Despite coming from a family of mathematicians, as a woman Noether had to fight against the odds to pursue her mathematical interests. Women were not allowed to enroll formally at the University of Erlangen at the time, so she audited the courses instead and aced the final exams anyhow. Later the university lifted restrictions on female students, but despite earning a doctorate with the highest honors, Noether had trouble finding a teaching position.

When Noether was being considered for appointment as a Privatdozent (the equivalent of an associate professor) at the

Emmy Noether (Science Photo Library)

University of Göttingen, one faculty member protested: "What will our soldiers think when they return to the university and find that they are required to learn at the feet of a woman?" The prominent mathematician David Hilbert, one of Noether's champions, was infuriated by this discrimination. "I do not see that the sex of the candidate is an argument against her . . . After all, we are a university, not a bathhouse," he argued. But he was unsuccessful at persuading his colleagues, and Noether was offered a guest lecturer position instead. With the rise of Hitler, she was among the first Jewish academics to be fired and was compelled to flee the country. She moved to the United States in 1933, to accept a professorship at Bryn Mawr College in Pennsylvania. Today a prestigious funding scheme for young scientists from the German Research Foundation is named in Noether's honor.

Soon after arriving at Göttingen, while investigating aspects of Einstein's just-published general theory of relativity, Noether made the explicit connection that symmetry implies a conservation law and vice versa. Where there is a symmetry or regularity in nature, she proposed, there is an associated law of conservation. For example, the fact that the laws of physics remain the same over time (or "invariant," to use the technical term) implies conservation of energy. Indeed, whether you drop a coin from the balcony tomorrow or three weeks from now, it will reach the ground with the same acceleration. Conversely, if nature obeys energy conservation, the laws of physics are symmetric with respect to time. Noether's realization was so profound that it is now referred to as Noether's theorem. As the Nobel Prize–winning physicist Leon Lederman and his Fermilab colleague Christopher Hill wrote in *Symmetry and the Beautiful Universe*, it is "one of the most important mathematical theorems ever proved in guiding the development of modern physics, possibly on a par with the Pythagorean theorem."

The reason that Lederman and Hill believed Noether's theorem was so significant has a lot to do with our understanding of nature at its deepest level, including the subatomic realm. Take a muon particle, for example, which emits an electron to the right when it decays. Given the rules of symmetry, an antimuon should emit the electron to the left. Physicists refer to this property as parity (or "handedness"), and as predicted by Noether's theorem, they expect particle interactions to conserve parity, along with other quantities such as charge and energy. However, the standard model allows these symmetries to break down occasionally, in a phenomenon called the charge-parity, or CP, violation. So, once in a while, an antimuon might decay to the right, breaking the usual rule.

Physicists James Cronin and Val Fitch, both then of Princeton University, and their colleagues observed CP violation for the first time in 1964 in a synchrotron at the Brookhaven National Laboratory on Long Island, New York. They noticed that electrically neutral particles called kaons, or K-mesons, can transform into their antiparticles and vice versa, but the transformation doesn't occur with the exact same probability in both directions. In essence, they found that nature was not quite evenhanded between matter and antimatter. The effect was tiny, though, and nowhere near enough to account for the dominance of matter over antimatter. Since then, physicists have been looking for other instances of CP violation that might boost the numbers. They began operating two of the most sensitive experiments that search for CP violation in 1999. One is named BaBar and is being conducted at the Stanford Linear Accelerator (SLAC) in California. The other is called Belle and is housed at the

KEK laboratory in Japan. These "B factories" smashed together electrons and positrons at speeds close to that of light to create a shower of neutral particles known as B mesons, which disintegrate in just a trillionth of a second into a variety of other exotic species. By recording billions of such decays over nearly a decade of operations, physicists have measured a larger-than-expected asymmetry in the decay rates of B-mesons compared to those of anti-B-mesons. This is the most severe case of CP violation yet seen, but it is still far from sufficient to account for the matter-antimatter asymmetry in the universe, and so the search continues for a more effective mechanism. Starting in 2011, one of the five major experiments at CERN's Large Hadron Collider, the world's most powerful particle accelerator, occupying a circular tunnel some 27 kilometers (16.8 miles) long near the Swiss-French border, is looking for other instances of CP violation.

Meanwhile, theorists have dreamed up a number of other ways to generate a stronger asymmetry. Some of these are rather complicated, if not contrived. Rocky Kolb describes the more outlandish suggestions as "constructing a grand edifice to explain this single number." One idea is that evaporation of primordial black holes, formed in the infant universe, did not conserve the symmetry. Another explanation looks more promising: perhaps the asymmetry has its origins in the family of particles known as leptons, which includes the neutrino. For this explanation to work, physicists have to assume that the early universe was filled with super-heavy counterparts to the light neutrinos observed today. When these super-heavy neutrinos decayed, they may have had a slightly higher probability of making matter than anti-matter. Since we do not have accelerators at present with sufficient energies to create these hefty particles, scientists have to rely on observing the properties of their lighter cousins to infer whether that could indeed have been the case.

The reason that neutrinos play a starring role in this cosmic mystery has to do with a strange property they *may* possess: neutrinos may turn out to be identical to antineutrinos. Most elementary particles have corresponding antimatter twins with opposite charge and spin. Since neutrinos and antineutrinos have no charge, the only way to distinguish them is by their spin: neutrinos spin left while antineutrinos spin right. Despite the difference in spin, both may interact with matter the exact same way. If that were true, and the two types of particle were perfectly interchangeable, that could help explain how matter came to predominate over antimatter in the early universe.

The Italian physicist Ettore Majorana, the reclusive genius who vanished without a trace at age thirty-two, was the first to suggest that neutrinos may possess a dual identity. Born in 1906 to a prominent Sicilian family, Majorana showed a knack for arithmetic and chess as a child. While studying in Rome to become an engineer, following in his father's footsteps, he made friends with Emilio Segrè, who persuaded him to switch to physics from engineering. Like Segrè himself, Majorana joined the research group of Enrico Fermi, who had just been appointed a professor at the University of Rome at the tender age of twenty-six. Fermi's group of ambitious young physicists came to be known as the "Via Panisperna boys," after the street where the physics institute was located.

Majorana's theoretical work focused on the structure and behavior of atoms and their constituents. A grant from the Italian research council allowed him to travel to Leipzig, Germany, to work with Werner Heisenberg, and to Copenhagen to collaborate

Ettore Majorana (E. Recami and
F. Majorana)

with Niels Bohr. While working in Germany, he developed a
serious case of gastritis, and his illness affected him for years,
even after his return to Rome. Perhaps as a result, he published
hardly any scientific papers, often dismissing his own work as
humdrum research not worthy of publication. Despite Fermi's
urging, he even failed to publish his prediction of the neutron's
existence, and so he was denied any credit when other scientists found it independently.

With Fermi's backing, which counted for a lot in Italy at the
time, Majorana was offered a chair in theoretical physics at the
University of Naples in late 1937. He moved into a hotel there
and began teaching at the university the following January. For
the first couple of months, his work seemed to go well. And
then, on March 23, he took the night boat from Naples to Pal-

ermo, on the island of Sicily, after withdrawing his savings. Two days later, he wrote a startling letter to the director of the physics institute in Naples: "I made a decision that has become unavoidable. There isn't a bit of selfishness in it, but I realize what trouble my sudden disappearance will cause you and the students. For this as well, I beg your forgiveness, but especially for betraying the trust, the sincere friendship and the sympathy you gave me over the past months . . ."

But Majorana apparently changed his mind soon after posting it, because he telegraphed a message asking his colleague to disregard the previous letter. As he wrote in a second note dated March 26: "The sea rejected me and I'll be back tomorrow at the Hotel Bologna traveling perhaps with this letter. However, I have the intention of giving up teaching." His worried colleague alerted Majorana's family. Majorana had bought a ticket in Palermo to take the boat back to Naples on the night of March 25, but he never showed up back on the mainland. Despite the offer of a monetary reward from his family, no other clues turned up. Fermi even appealed to the Italian prime minster Benito Mussolini to support a search: "I have no hesitation to state to you, and I am not saying this as an hyperbolic statement, that of all Italian and foreign scholars that I have had the opportunity to meet, Majorana is among all of them the one that has most struck me for his deep brilliance."

Speculations about Majorana's fate range from the obvious to the sinister. The plain possibility is that he committed suicide, jumping off the boat into the Tyrrhenian Sea. But if that were his plan, why would he take out his savings from the bank? In addition, his family members insisted that suicide would run contrary to his strong Catholic faith. Some people thought that Majorana had suffered a spiritual crisis, and joined a monastery.

One Jesuit priest reported being approached about joining the order by a distraught young man of Majorana's appearance, while monks at a monastery south of Naples also claimed to have seen him. Yet another theory is that he absconded to Argentina. Reports of Majorana sightings in South America continued for decades after his disappearance. One Sicilian writer has argued that Majorana went into hiding because he foresaw the advent of nuclear weapons and did not want to have anything to do with this terrible development. Others have suggested that he was killed by Nazi agents or by the Sicilian Mafia.

For all his brilliance, Majorana may have had trouble coping with life's little frustrations. As Fermi put it, "Majorana had greater gifts than anyone else in the world. Unfortunately he lacked one quality which other men generally have: plain common sense." His legend lives on in the pop culture of his homeland, not only through conspiracy theories and Majorana sightings (which were supposedly as common in Italy at one time as Elvis sightings in America) but also as a comic book hero and a ubiquitous trademark.

Meanwhile, his stock in the scientific world has risen dramatically in recent years. That's because of growing interest in a suggestion he made about the nature of neutrinos in a paper published the year before his disappearance. As I mentioned earlier in this chapter, Dirac's equation describing the electron, which has a negative electric charge, predicted the existence of a positively charged antimatter twin. However, Majorana realized that the chargeless neutrino could be its own antiparticle. There need not be a mirror twin of it in nature. Thus Majorana was able to write down a simpler equation to describe the neutrino than Dirac's for the electron. It is said that he was reluctant to publish his result, as usual, despite Fermi's prodding. Some say that Fermi himself wrote the paper, based on Majorana's notes, and submitted it for publication under Majorana's name. If not for this generous act on Fermi's part, we might never know about Majorana's pivotal insight.

If Majorana's prediction turns out to hold water, and indeed the neutrino is its own antiparticle, there are incredible implications for physics. For one, the standard model, the ultimate rule book for the subatomic world, would have to be revised. For another, the dual nature of the neutrino could be central to the symmetry breaking that resulted in matter winning over antimatter, and could thus help account for our very existence. Notwithstanding its apparent simplicity, Majorana's hypothesis is a tough one to test through experiments. It is only now that scientists are getting ready for rigorous trials, as we will see later in this chapter.

As you may recall from chapter 2, Wolfgang Pauli first invoked the neutrino to account for the energy that seemed to vanish during the so-called beta decay of a radioactive nucleus. Specifically, when a neutron decays into a proton and an electron, an antineutrino is also produced. This decay process conserves electric charge: you start with a neutron, which has no charge, and end with a +1 proton, a −1 electron, and a chargeless antineutrino, thus a net charge of zero again. It also conserves the lepton number: it is zero at the beginning and also at the end, because the electron has a +1 lepton number while the antineutrino is assigned −1. So far, so good: we are playing by the rules of the standard model.

As it turns out, a rare variation of beta decay is the key to testing Majorana's hypothesis. Already back in 1935, the German-American physicist Maria Goeppert-Mayer conceived of an unusual form of "double beta decay": two neutrons in the same nucleus could decay into two protons simultaneously, releasing two electrons and two antineutrinos. This process too would conserve both electric charge and lepton number. Her calculations showed that such an occurrence would be quite rare. Besides, it would be rather difficult to observe, because the much more common single beta decays would swamp the signal. After decades of experimental efforts by many physicists, Michael Moe of the University of California, Irvine, and his colleagues finally observed double beta decay in the laboratory in 1987.

Long before Moe's team detected double beta decay experimentally, scientists realized that an even weirder and rarer type of beta decay is also possible if neutrinos are their own antiparticles, as Majorana had suggested. If Majorana was onto something, then two neutrons can undergo beta decay together such that the antineutrino emitted by one is immediately absorbed by the other. The net result is two neutrons disintegrating simultaneously without releasing any antineutrinos or neutrinos. Physicists have given it the stodgy name "neutrinoless double beta decay." This process would conserve electric charge, but not lepton number: you would start with zero but end up with +2 from the two electrons and no antineutrinos to balance the budget. That means neutrinoless double beta decay would violate an important law of conservation, a cardinal rule of the standard model, and could thus provide a mechanism for breaking the symmetry between matter and antimatter.

If scientists do observe this exceedingly rare process in nature, it would prove Majorana's hunch: that the neutrino is its own antiparticle. It would represent a momentous breakthrough as the first-ever case of lepton number nonconservation, and in turn pave the way to understanding how matter came to dominate the universe. What's more, the probability of neutrinoless double beta decay is closely tied to the mass of the neutrino: the larger the neutrino mass, the higher the decay rate, so measuring the rate of this rare process will tell scientists the mass of the neutrino directly. So far, physicists have only been able to measure the mass differences between neutrino types (as we saw in chapter 5), not their absolute mass. In other words, measuring this rare process would achieve two goals at once—we'd find out the mass of the neutrino and whether the neutrino is its own antiparticle.

The needle-in-a-haystack search for neutrinoless double beta decay is heating up, as physicists build sophisticated experiments to put Majorana's hypothesis to the test, more than seventy-five years after he made the proposal. The fact that it has taken this long to build experiments with sufficient sensitivity shows, says Giorgio Gratta of Stanford University, "how difficult it is to measure this property, but also how brilliant Majorana was."

Over the next five years, several sensitive experiments are expected to begin operations. One of them is called the Cryogenic Underground Observatory for Rare Events (CUORE), and it is being built at the Gran Sasso laboratory in Italy. It uses some 200 kilograms (440 pounds) of tellurium, about a third of which is radioactive and undergoes double beta decay. CUORE's detectors are cooled down to a smidgen above absolute zero (which makes it easy to record even a small rise in temperature due to the absorption of a particle or a gamma ray). Since neutrinoless double beta decay would occur extremely rarely, if it occurs at all, scientists have to suppress other sources of noise that would otherwise overwhelm its signal. The first step is locating the experiment within Gran Sasso Mountain in the Apennines, beneath a mile of rock that would block most cosmic rays bombarding the Earth from space. Second, the detectors are assembled in ultra-clean facilities, to prevent contamination from common radio-active elements in the environment. Third, a 3-centimeter-thick (1.2-inch) lead lining will surround the detectors, to shield them from any radioactivity in the rock.

Newly mined lead is itself slightly radioactive, though, which means its own emissions would interfere with the beta decay measurements. So the physicists designing the CUORE experiment wanted to use very old lead that had lost almost all traces of natural radioactivity over time. And they went to ex-traordinary lengths to secure the perfect shielding material: the metal they use comes from a cargo vessel that sank off the coast of Sardinia 2,000 years ago. A scuba diver discovered the ship's remains in 1988, including over a thousand ingots of lead, prob-ably intended to make slingshot ammunition for Roman soldiers. Archaeologists wanted to study the trademarks on the ingots for insights into ancient maritime trade, but did not have enough funds to recover them all. That's when physicists stepped in, and Italy's National Institute of Nuclear Physics contributed the equivalent of about $200,000 toward the salvage operation. In return, physicists received a fraction of the recovered bounty, a total of almost ten tons of the worst-preserved ingots, which will be melted down to make the lining for the CUORE experiment.

The CUORE team is in friendly competition with a few other groups chasing neutrinoless double beta decay. Another European group is already operating an apparatus that looks for the exotic process in a block of germanium at the same Gran Sasso laboratory. American scientists have begun their own ex-periment called EXO-200 (Enriched Xenon Observatory), which uses 200 kilograms (440 pounds) of liquid xenon. It is located under a salt bed near Carlsbad, New Mexico, that also serves as a nuclear waste depository. The same salt layer that shields the nuclear waste also protects the EXO-200 detector from cosmic rays and radioactivity in the rocks. The xenon is contained in a copper drum, which is in turn surrounded by a solvent that acts like antifreeze to keep it at just the right temperature. The researchers took extreme care to ensure that their apparatus is free of background radiation that could hamper the hunt. They assembled the EXO-200 experiment inside a large clean room in a facility with a thick concrete roof at Stanford University, using materials free of radioactivity and tools cleansed with acetone and alcohol. Since flying would expose the equipment to more cosmic rays, they transported it in sealed containers by truck from California to New Mexico, a distance of 1,300 miles. To further reduce exposure to cosmic rays, they tried to limit the time spent on the road by employing two drivers to take turns at the wheel so that the vehicle would not have to stop for rest breaks. The researchers went so far as to paint the main container with a special reflecting paint to keep it cool while on the road, and they used a truck equipped with air-ride suspension to prevent vibrations from damaging their delicate instruments. Meanwhile, Japanese physicists have also begun a search for neutrinoless double beta decay at the Kamioka mine, using 400 kilograms (880 pounds) of xenon contained in a giant nylon balloon.

In fact, a decade ago, a small group of physicists led by Hans Klapdor-Kleingrothaus at the Max Planck Institute for Nuclear Physics in Heidelberg, Germany, claimed to have detected neutrinoless double beta decay. They analyzed many years of data from a joint Heidelberg-Moscow experiment, which used five large, ultrapure crystals of enriched germanium located at the Gran Sasso underground laboratory, and reported evidence of this rare transformation. However, other researchers, including those from the Moscow side of the collaboration, have pointed to flaws in their analysis and questioned whether the measurement is a statistical fluke. As Hitoshi Murayama says, "Most theories don't predict as large rates as they claimed to see." Stanford's Gratta, who is a key player in the EXO-200 experiment, agrees: "The community consensus is that neutrinoless double beta decay has not been seen yet." If by any chance Klapdor-Kleingrothaus and his colleagues are right in their contention, other experiments should be able to validate their finding within the next few years.

"Since it is such a profound measurement, the physics community agrees that we need independent results from multiple experiments to be sure," says Karsten Heeger of the University of Wisconsin–Madison, who is a member of the CUORE collaboration. "If we see something, it would be a very big deal," he adds. There are several reasons why the detection of neutrinoless double beta decay would shake up physics, astronomy, and cosmology. First, detecting this decay would prove that Majorana was right, and that neutrinos are indeed their own antiparticles. Second, physicists would have a direct measure of the absolute mass of neutrinos, something that has eluded them for decades, and so astronomers would learn whether these particles were massive enough to play a critical role in forming the first clumps of matter in the early universe. Third, the decay would violate the conservation of lepton number, and, as Heeger puts it, would "indicate breaking of a fundamental symmetry in physics, and thus require a major revision in the standard model." Fourth, it would help cosmologists understand how matter came to predominate over antimatter seconds after the big bang. Given all these exciting prospects, it is no wonder that neutrino hunters are looking forward to the coming decade with great anticipation.

EIGHT

SEEDS OF A REVOLUTION

The summer of 2012 marked a triumphant capstone for physics. Two separate experiments at the gigantic Large Hadron Collider (LHC) at the CERN laboratory revealed compelling evidence of the Higgs boson, one of the most elusive subatomic particles that theorists had ever concocted. With this discovery, the crucial final piece of the grand edifice known as the standard model of particle physics fell into place.

But the curious antics of neutrinos threaten to bring down the physicists' elaborate creation, or at least show that it is incomplete. Physicists reckon that the finding that neutrinos have some mass, as minuscule as it is, requires them to modify the standard model. Neutrino hunters have begun to see hints of phenomena that may require a major overhaul. As they track down the vagaries of neutrino behavior with powerful new experiments, these scientists will not only advance our understanding of the fundamental nature of matter but will also shed light on the first crucial seconds after the big bang and on the fiery death throes of stars. Along the way, they hope to use neutrinos to probe the Earth's internal heat source, pinpoint mineral deposits underground, and even help prevent nuclear proliferation.

What's more, they promise to deliver all this for a reasonable price to the taxpayers who bear most of the costs of doing basic science.

Chasing the Higgs boson, on the other hand, cost billions of dollars and took several decades. The Higgs hunt began innocuously enough, with a proposal from six physicists, working in three independent teams, in the early 1960s. They suggested that an invisible force field permeating space was responsible for endowing some elementary particles with mass. As is often the case in fundamental physics, their suggestion was motivated by mathematical considerations of symmetry in nature. The proposed force field has come to be known as the Higgs field, after Peter Higgs of the University of Edinburgh, one of the six theorists who proposed it, and it is an essential feature of the standard model.

The most direct way to test whether the Higgs field exists and to determine its properties is to detect the particle associated with it. In quantum mechanics we think of the Higgs boson as a vibration in the Higgs field. If there were no field, there would not be any vibrations, so detecting a particle would prove the corresponding field's existence. To create a vibration in the field, you have to disturb it, much as you would disturb a pond if you dropped a pebble into it. Scientists hoped that if they collided particles with enough energy in a particle accelerator, that would create a strong enough perturbation in the Higgs field to observe the Higgs boson. Unfortunately, the Higgs field theory provided little guidance to experimentalists: it did not specify the mass of the Higgs boson, so they did not know how energetic the collisions would have to be for it be detectable. Some prominent scientists were skeptical that it would ever be identified. Stephen Hawking, for one, wagered a hundred dollars against Gordon Kane of the University of Michigan, a proponent of Higgs, that the particle would not be found.

Finding the Higgs boson, or ruling out its existence, was a top priority for the LHC, built over a decade at a cost of nearly $9 billion with the help of thousands of scientists and engineers. Not surprisingly, when CERN scheduled a press conference on July 4, 2012, many people expected it to announce the discovery of this long-sought particle. Hundreds of people lined up overnight to get into the auditorium. Journalists reported that the atmosphere at the laboratory was reminiscent of a rock concert. At the event, scientists presented results from two LHC experiments, showing significant peaks above the noise in each data set. The peaks appeared at an energy of about 125 GeV (giga–electron volts), corresponding to a particle about 130 times more massive than the proton. The researchers had

The ATLAS detector, one of the two experiments at the Large Hadron Collider that found evidence for the Higgs boson (CERN)

little doubt the bumps signaled the discovery of the Higgs boson.

Peter Higgs, who was in his eighties by this point, was a guest of honor at the announcement along with two other theorists who had predicted the particle's existence, and attendees witnessed him wiping away a tear of joy. "It's certainly been a long wait," he said at a press conference in Edinburgh a couple of days later. "At the beginning I had no idea whether a discovery would be made in my lifetime, because we knew so little at the beginning about where this particle might be in mass, and therefore how high energy machines would have to go before it could be discovered," he added. Stephen Hawking paid up his bet with Gordon Kane. Like many other physicists, Hawking agreed that tracking down the Higgs boson was a major milestone in the history of physics. However, in an interview with the BBC, he also noted the flip side of what the discovery meant: "But it is a pity in a way because the great advances in physics have come from experiments that gave results we didn't expect."

New data from the LHC, since the initial announcement, have firmed up the Higgs detection beyond a reasonable doubt. Scientists are scrutinizing the data for the slightest anomalies that could hint at unexpected phenomena, like the possibility that more than one Higgs particle exists. However, as Hawking lamented, while finding Higgs in some sense marks the culmination of a long, arduous, and enthralling journey, it does not point the way to the next great scientific adventure, because so far it has behaved as expected. As Steven Weinberg of the University of Texas at Austin, one of the architects of the standard model and a Nobel laureate, put it to me, "Higgs is the last missing piece of the standard model, but it doesn't take us beyond."

For that, more and more physicists are looking anew at the elusive neutrinos, because they do show hints of physics beyond what we've explored so far. For example, the standard model in its original form assumes that neutrinos have no mass. Thus, as I discussed in chapter 5, the finding that they flip among three flavors and must possess a small but nonzero mass came as a surprise. As Weinberg explained when commenting on the mass of neutrinos, "It's the only thing we have discovered in elementary particle physics that gives definite evidence of something beyond the standard model. But it is a clue we haven't been able to interpret yet."

Georg Raffelt of the Max Planck Institute for Physics in Munich, Germany, shared a similar sentiment. "The zero mass of neutrinos was taken as a self-evident truth by theorists. Nature has straightened us out," he told me when we discussed evidence for physics beyond the standard model. "The discovery of neutrino flavor oscillations has truly opened a new frontier, a new direction of research to chase," he added. That explains in large part why neutrino physics has transformed from a sleepy backwater twenty years ago, when only a handful of scientists paid any attention, to a thriving hub of activity, with over a thousand researchers actively studying these shadowy particles.

Physicists have placed limits on the mass of neutrinos, but haven't measured it directly yet. The data they have so far suggest that a neutrino is at least a million times lighter than an electron, which itself is puzzling in the context of the standard model, according to André de Gouvêa of Northwestern University. As he puts it, "Zero we could understand, but a very small mass is surprising." Theorists have invoked an extension of the standard model, called the "seesaw mechanism," to explain how neutrino masses came to be so much smaller than those of other elementary particles. But for this process to work, there must be very heavy counterparts to the light neutrinos that we're familiar with. It may be that such hefty neutrinos existed right after the big bang, when the universe was much hotter and denser than it is now, and decayed into other particles quickly.

While the hypothetical heavy neutrinos remain beyond our experimental reach, physicists hope to nail down the absolute masses of their lighter cousins over the next decade. One experiment in Germany, dubbed KATRIN (for Karlsruhe Tritium Neutrino Experiment), will attempt to do so by making precise observations of beta decay, the same process that led Wolfgang Pauli to invoke the existence of neutrinos in the first place. A giant spectrometer attached to KATRIN will measure the energies of electrons released during the radioactive decay of tritium (a heavy form of hydrogen that contains two neutrons and a proton), and infer the masses of the antineutrinos that flee the scene.

After the technicians assembled the 200-ton spectrometer for KATRIN at a plant 250 miles from the experiment site in Karlsruhe, they faced an unusual stumbling block: the device was too big for the region's narrow roads. So they put it on a barge and floated it on the Danube River through Austria, Hungary, and Serbia into Romania, where they transferred it to a ship for crossing the Black Sea and the Aegean. By the time the cargo arrived in Sicily, a storm had blown away its protective cover. At a Sicilian port, workers loaded the device onto a heavy lift vessel that would carry it through the Mediterranean Sea, through the Strait of Gibraltar, around Spain and France to the mouth of the Rhine River in northern Germany. Since the Rhine's water levels were low, the ship couldn't enter it, so the workers

The main spectrometer of the KATRIN arrives in the village of Leopold-shafen, near Karlsruhe in northwestern Germany, after a 5,600-mile journey (Karlsruhe Institute of Technology)

transferred the spectrometer onto a pontoon for the next leg. Later they used a crane to offload the device onto a truck, which squeaked it through the village of Leopoldshafen with great fanfare to its destination, where it arrived after a 5,600-mile circuitous journey that took two months. Once technicians complete assembling and testing the equipment, KATRIN is expected to begin taking data by 2015.

One other reason for the skyrocketing interest in neutrinos has to do with their crucial role in bridging disparate fields of science. Besides the promise of advancing fundamental physics and hinting at the need for a theory beyond the standard model, neutrinos are pivotal for cosmology. They could reveal how the universe came to be dominated by matter over antimatter, as we discussed in the last chapter, and could also help us understand the growth of large-scale cosmic structures such as clusters of galaxies. In fact, one of the best limits on the absolute mass of the neutrino comes from comparing the distribution of galaxies in space to the pattern of ripples in the big bang's afterglow called the cosmic microwave background. According to Licia Verde of the University of Barcelona in Spain, future sky surveys offer our best hope for pinning down the neutrino mass. "If the total mass is below 0.2 electron volts . . . then no planned neutrino experiment can determine the neutrino mass in a model-independent way," she explains. So instead of relying on Earth-bound experiments, we may have to look upward. As Verde says, "Planned cosmological surveys have enough statistical power to see in the sky the effect of a neutrino mass as small as the minimum allowed by oscillations."

On the flip side, neutrinos could in principle allow us to probe the universe right after it came into being. For now, the farthest we can look back is about 380,000 years after the big bang, as the seething primordial soup of particles that made up the infant universe was opaque to light (and other electromagnetic radiation) before that time. Neutrinos, on the other hand, could reach us from a much earlier epoch, because they could have traveled freely without interacting with ordinary matter since just a couple of seconds after the birth of the universe. These relic neutrinos should still be all around us, averaging about 150 in each cubic centimeter of space (or 2,500 per cubic inch), but they would have very low energies by now, and we haven't figured out how to detect them yet. Some physicists hope that future experiments will be able to register a handful of relic neutrinos per year, though they will need many more to make useful measurements.

In the meantime, neutrinos matter to astrophysics because

their characteristics determine the life cycles of stars, the production of heavy elements through fusion, and the spectacular explosions of massive stars as supernovae, which help spread the materials essential for life. They are central to nuclear science because nuclear power generators and nuclear bombs produce staggering numbers of these particles.

Even geophysicists are looking to neutrinos. They hope that neutrinos will help settle a long-standing question about the sources of the Earth's internal heat, one that triggered a dispute between the naturalist Charles Darwin and the physicist William Thomson (later Lord Kelvin) some 150 years ago. In the first edition of *On the Origin of Species by Means of Natural Selection*, published in 1859, Darwin estimated that it would have taken roughly 300 million years to erode a great valley in the south of England, so the Earth had to be at least that old. Thus, he argued, the Earth had been around long enough for the process of evolution to lead to the remarkable diversity of life on our planet. Thomson, one of the foremost scientific luminaries of the time and a critic of natural selection, doubted Darwin's claim. So he derived his own independent estimates for the age of the Earth, by assuming that the Earth had cooled steadily from a molten initial state, and for the age of the Sun, assuming that it was powered by slow contraction under its own gravity. Both of Thomson's calculations resulted in relatively young ages that scientists deemed too short to account for biological evolution. Darwin worried about the discrepancy between his estimates and Thomson's age estimates. In a letter to Alfred Russel Wallace, the codiscoverer of natural selection, Darwin complained that "Thomson's views on the recent age of the world have been for some time one of my sorest troubles." Darwin went so far as to remove any mention of specific timescales from later editions of his tome.

Back then, scientists were not aware of nuclear fusion, which we now know fuels the Sun, and they had not discovered radioactivity, which provides a continuing heat source inside the Earth. Thomson's age estimates, therefore, came in low because he didn't account for these energy sources. Now the age of the Earth and the solar system is well established through radiometric dating of primitive rocks and meteorites, so there is no longer a mismatch: 4.5 billion years allows plenty of time for enormous geological changes and dramatic biological evolution.

However, there is still some uncertainty among geophysicists about just what fraction of the Earth's heat is generated through nuclear fission of radioactive elements and what portion is left over from its fiery birth. They believe that the decay of uranium and thorium accounts for most of our planet's energy production, but aren't sure about the abundance of these elements or how they are distributed through the Earth's interior. As it happens, we know the Sun's composition much better than the Earth's.

Now, thanks to sensitive modern detectors, scientists can probe our planet's interior by observing the antineutrinos produced during the decay of radioactive elements inside the Earth. While the technical capability is new, the basic idea is not. As far back as 1953, in a letter to Frederick Reines, physicist and cosmologist George Gamow pointed out the possibility of detecting neutrinos of terrestrial origin. Then, in the early 1980s, three other physicists—Lawrence Krauss, Sheldon Glashow, and David Schramm—worked out in detail how many neutrinos should emerge from the Earth each second. "Emboldened by the remarkable experimental detection of solar neutrinos . . .

several colleagues and I started to think about other natural sources of neutrinos. One was right below our feet, literally," Krauss recounted. "When we calculated how many such antineutrinos might be produced by all the radioactive materials thought to be in the earth, the number was almost as large as the solar neutrino flux across a small energy range," he added. "We also realized that it would be much harder than it had been for Davis to detect solar neutrinos (which was plenty hard). So we wrote up the paper, figuring such a study would never be done." The trio demonstrated that in principle, antineutrinos would allow us to peer deep into the Earth's bowels and would tell us a lot about its makeup, because the number of particles emitted each second is a direct measure of the planet's total radioactivity. So observing these neutrinos produced inside the Earth, now called geoneutrinos, would reveal the amount of uranium and thorium in the Earth's crust and the mantle below it.

Flash forward two decades to Japan. In 2005, the KamLAND neutrino detector, originally built for particle physics experiments, spotted the geoneutrinos that Gamow had predicted. The total number recorded was a mere twenty-five, but it was an important step. As the KamLAND collaboration's spokesperson Atsuto Suzuki announced in a press release about the progress, "We now have a diagnostic tool for the Earth's interior in our hands. For the first time we can say that neutrinos have a practical interest in other fields of science."

It took five years before the Borexino experiment in Italy provided independent confirmation of what the Japanese team had reported. As of 2011, the KamLAND team has refined their statistics, recording more than 100 geoneutrinos, thanks to improvements in their detector. They were also helped by the unexpected shutdown of nearby nuclear reactors, which otherwise produce lots of neutrinos of their own and complicate the measurements of geoneutrinos.

Based on the number of geoneutrinos that they've detected, scientists infer that the decay of radioactive elements inside the Earth produces about 20 terawatts of heat. (For the sake of comparison, humanity's total power consumption adds up to about 15 terawatts.) Separately, they can estimate how much heat the Earth releases altogether, by measuring temperature at the bottom of boreholes. That number comes to about 44 terawatts. That means radioactive decay is responsible for less than half the total. The rest is heat from the planet formation process, still trapped in the Earth's interior. These findings rule out geophysical models that assume the Earth has lost all its primordial heat and depends entirely on radioactivity for internal energy.

With more precise measurements, scientists will be able to learn about the composition of the Earth's building blocks and test scenarios for their assembly. They also hope to learn what role, if any, the heat from radioactivity plays in driving plate tectonics, the slow movement of chunks of the Earth's crust that is responsible for shaping continents and mountains. As Georg Raffelt pointed out, "Bore holes, earthquakes, and volcanoes give us clues about the upper layers, but neutrinos could probe much deeper. They could tell us whether we have modeled the Earth correctly."

Soon a geoneutrino detector will begin operations at the Sudbury Neutrino Observatory in northern Ontario, Canada. Referred to as SNO+, it will be the deepest, at a mile and a quarter beneath the surface, and the most sensitive yet. The depth helps cut down the "noise" caused by cosmic rays from space that interfere with neutrino detection. SNO+ also happens to be located far from any nuclear plants, so, unlike Kam-

LAND, it will not have to contend with a flood of neutrinos produced in nearby reactors swamping the already faint geoneutrino signal. Other detectors that have been proposed, such as LENA (for Low-Energy Neutrino Astronomy) in Europe and INO (for India-based Neutrino Observatory), will also dabble in geoneutrino hunting. In addition, John Learned of the University of Hawaii has been pushing for an antineutrino observatory on the floor of the Pacific Ocean. Placed above the thin ocean crust, it could measure the mantle's contribution to the Earth's heat better than land-based detectors. Current models assume that uranium and thorium are distributed uniformly throughout the mantle, but according to Learned, that may not be the case. For example, he notes, these elements may have piled up at the boundary between the Earth's core and the mantle. Data from various locations around the world would allow geophysicists to map their distribution in order to better understand the origin of the Earth's radioactivity.

Some geophysicists wish for even greater rewards. Since neutrino oscillations are sensitive to density variations along their path, these researchers want to send neutrino beams from one point on the Earth to another through the crust in order to probe its structure. If they could use intense beams of neutrinos that are produced in accelerators and highly sensitive detectors, they would be able to scan the Earth's crust and look for big cavities that may be filled with water or mineral deposits, just as dentists use X-rays to scan your teeth to search for cavities. The technique may even uncover geological formations likely to contain oil, which would be of great interest to companies seeking new petroleum reserves. For this scheme to work, though, the neutrino beam would need to be thousands of times more intense than what particle accelerators are able to produce currently. For now, oil prospecting with neutrinos remains impractical.

Meanwhile, in a delightful stroke of scientific serendipity, neutrino hunters have lent a helping hand already to marine biologists interested in deep-sea marine life. The unlikely collaboration came about because a pesky noise for one turns out to be a precious signal for the other. In 2005, Italian physicists building the Neutrino Mediterranean Observatory (NEMO) off the coast of Sicily looked into how they might "listen" to the particles, rather than simply record the faint flickers of light resulting from the occasional interactions between neutrinos and water molecules. They were motivated by theoretical calculations showing that a high-energy neutrino interacting with a molecule should also trigger a puny sound wave. Since sound travels farther than light underwater, deploying acoustic detectors could increase the odds of capturing neutrino hits. In fact, scientists reckon that the telltale "pop" caused by a neutrino could be heard from a few miles away with acoustic sensors. The challenge, of course, is teasing it out from other noises in the surroundings.

The NEMO physicists had no clue what sounds would dominate the Neptunian abyss a mile or more beneath the surface of the Mediterranean Sea. Marine biologists told them the deep waters were unlikely to be silent, but they did not have the data to be sure. So they welcomed any help from the physicists to find out. In early 2005, the two groups of scientists set up four ultrasensitive hydrophones (microphones designed for underwater use) at the NEMO test site. The devices were hooked to a cable that relayed data to a computer on a nearby pier. Predictably, the hydrophones recorded ambient sounds from natural water movements and maritime traffic. There were occasional bursts caused by a large ship's propeller as well. But what grabbed the scientists' attention were the booming clicks made by sperm whales as they compressed air through the cavities in their nasal passages. Among the loudest sounds made by an animal, these clicks apparently help the whales gauge the depth of the ocean and help them locate prey, the same way bats use sonar for navigation. The researchers also heard sequences of clicks, called codas, which whales use to communicate with each other.

After listening to hundreds of hours of recordings, the biologists concluded that sperm whales are much more common in the Mediterranean basin than previous studies closer to the surface had suggested. Now they want to use the deep-sea acoustic system to track the numbers and movements of the cetacean population over time. They hope to learn whether sperm whales migrate from the Atlantic Ocean into the Mediterranean and back as the seasons change. Moreover, the initial findings have led to the deployment of acoustic equipment at several other sites around the world, including the ANTARES undersea neutrino telescope off the French coast.*

It is not clear yet whether physicists will be able to detect neutrinos with acoustic equipment, but plans for the next-generation deep-sea neutrino observatory, a multi-cubic-kilometer network called KM3NeT, include an array of hydrophones anyway. These devices help position the optical detectors, and they will come in handy for biologists keen to study not only sperm whales but also fin and beaked whales—all at unprecedented depths. In the meantime, NEMO researchers have stumbled upon another surprising discovery: chains of marine

---

* You can listen to a sampling of the sound recordings online at http://listentothedeep .com/acoustics/index.html.

vortices that oceanographers had not expected to see in the closed basin of the Mediterranean. Now they are trying to discern whether these vortices formed locally or traveled from hundreds of miles away. At one time simply the stuff of theoretical interest, neutrino hunting can now lead us to hidden treasures in unexpected ways.

Someday neutrino hunters may even contribute to world peace. Their exploits could help expose rogue reactor operators, catch dangerous plutonium smugglers, and even stop would-be nuclear bomb makers in their tracks. Scientists are now investigating the prospects for using neutrino detectors to prevent nuclear proliferation. As Georg Raffelt emphasized in this context, "Something as esoteric as neutrinos may turn out to have a practical use." Nuclear reactors used for power generation offer a potential source of weapons-grade plutonium, which builds up over time as uranium undergoes a fission reaction and splits into lighter elements. In its efforts to prevent the spread of nuclear weapons, the International Atomic Energy Agency monitors and inspects reactors in civilian use, and periodically, the IAEA inspectors compare the records kept by the reactor operators with their own monitoring data to assess whether there has been clandestine activity, such as frequent shutdowns to shuffle the fuel rods more often than necessary. Their current monitoring instruments need to tap into the reactor's plumbing— to keep track of the amount of coolant used, for example. But this equipment is unwieldy, expensive, and susceptible to tampering.

Fortunately, the same fission reactions that produce plutonium from uranium also release a by-product: antineutrinos. Detecting these antineutrinos would provide a direct, real-time measure of the nuclear reactions, thus a more reliable probe for the international monitors. As Raffelt put it, "Antineutrinos don't lie." There is no way to prevent these particles from escaping the reactor and revealing its activity if a detector could be placed nearby. Normal power plants burn the fuel rods continuously until the fissile material pretty much runs out, typically in about eighteen months. Over that time, as the fissile material is used up, the plant's power output—and the antineutrino flux—drops slowly and predictably. But if someone were interested in stockpiling plutonium, he or she would have to shut down the reactor for at least a day every few weeks to swap the fuel rods. "You need to bake the rods just the right amount to get plutonium," explains John Learned. "So if there is a reactor that shuts down once a month, then you know for sure they are cooking bomb material," he warns.

In theory, measuring antineutrinos coming out of a reactor would be a great way to keep tabs on the reactor's operations. In practice, however, there are a few complications. One of them is that it's difficult to build a neutrino detector that is relatively compact but sufficiently sensitive. The other is figuring out a way to shield it from stray particles such as cosmic rays. Besides, as Learned explains, "commercial power plant operators don't really want goofball physicists messing around their facilities. Who knows what trouble they might cause!" Even so, researchers from Lawrence Livermore National Laboratory and Sandia National Laboratories have tested a prototype at a nuclear power station in southern California. Their refrigerator-size instrument contained more than 1,300 pounds of mineral oil. The detector, placed 10 meters (33 feet) below the ground, was able to measure neutrinos from the nearby reactor and determine its power output to within a few percentage points of accuracy. It could tell within mere *hours* when the reactor was shut down.

While this test run produced promising results, it also pointed to the limitations of current technology. For one, mineral oil–based neutrino detectors are difficult to deploy. So researchers in several countries, including Brazil, Canada, France, and the United States, are experimenting with detectors that use water or plastic instead. In addition, placing a detector underground is not possible at all nuclear sites. But that is not essential, at least for monitoring the more powerful reactors. If the detector is close enough, the neutrino flux from the reactor would shine much more brightly than the random background events caused by cosmic rays. Some scientists expect that compact, reliable antineutrino detectors will be placed at cooperating nuclear reactor sites in the near future to verify that they adhere to the IAEA safeguards. There may even be an economic benefit to the power plant operators: they can improve the efficiency of their reactors by making appropriate adjustments based on real-time feedback from neutrino measurements. In the unfortunate event of an accident, neutrino detectors could trigger a timely shutdown of the reactor.

Other scientists have grander visions, which may be of interest to intelligence agencies and national security officials. They are investigating ways not only to monitor known nuclear sites from a distance but also to uncover clandestine reactors that have not been reported to the IAEA. Indeed, the most powerful reactors are difficult to hide, and since they emit a lot of heat, infrared satellites can spot them from space. Nuclear monitors are most worried about medium-scale facilities in rogue nations, which could be concealed more easily yet produce enough plutonium to make a bomb in only a year.

Based on initial calculations, Learned says that future technology should make it possible to detect nuclear sites remotely.

"Of course the farther away you are, the bigger the detector you need," he points out. The problem, however, is that bigger detectors suffer from noisier backgrounds, so they need to be buried underwater or underground. He envisions "a mobile underwater detector, perhaps mounted in a giant submarine, which could be towed into place." Such a contraption could remain in international waters and monitor a suspected country for stealth nuclear reactors from a safe distance. Learned even looked into acquiring an old Russian submarine to test this idea. In a similar vein, Thierry Lasserre of the French Alternative Energies and Atomic Energy Commission and his colleagues have proposed turning an oil supertanker into a mighty neutrino detector to look for undeclared nuclear sites. They have given their concept a name worthy of a spy novel: SNIF, for Secret Neutrino Interactions Finder. Both schemes sound far-fetched for now, given the formidable technological and political hurdles that need to be overcome, but they may be worthy of further investigation.

Someday we may also use neutrino detectors to watch over covert nuclear bomb tests. Current surveillance depends on techniques such as monitoring tremors in the earth. "They have missed some and have also had some false alarms, which turned out to be seismic events rather than bombs," according to Learned. Moreover, the use of cushioning materials and variations in cavity size can make it difficult to work out the exact parameters of a test explosion. "Detecting even one neutrino would tell you a great deal. If its timing matches a seismic event perfectly, then you know it was indeed a nuclear detonation," he explains. "If we measure ten neutrinos, we could nail down its magnitude pretty well." He suggests that a network of large neutrino detectors spread around the world could enhance surveillance of nuclear tests. Learned advocates for dual-use neutrino observatories that do fundamental science in addition to surveillance work. "The real bait for me, of course, is having these big detectors to do science," he admits.

One of the wackier ideas for putting neutrinos to practical use involves employing them for long-distance communication, since they travel unhindered through virtually anything. You could imagine sending an encoded beam of neutrinos from one side of the Earth to the other straight through the planet, whereas a radio signal needs to be bounced off multiple satellites in orbit or transmitted through undersea cables circling the globe. Along similar lines, one scientist has suggested using neutrino beams to send messages to submarines deep underwater. To have any chance of success, though, the beam would have to be about a million times more intense than those used in current experiments, and there would be some daunting challenges at the receiving end.

Nevertheless, a group of physicists have taken the first tentative step in neutrino communication. They have used a neutrino beam produced at Fermilab to fire pulses of these particles toward a giant underground detector more than half a mile away. The team relayed the word "neutrino" in standard binary code, which turns letters into strings of zeros and ones. The detector at the other end registered the simple message successfully, after it had traveled through a thousand feet of solid rock. The bit rate was a pathetic 0.1 bit per second, and it took over two hours to transmit the eight characters, but as the researchers wrote, "This result illustrates the feasibility, but also shows the significant improvements in neutrino beams and detectors required for practical applications." In other words, we shouldn't expect a neutrino phone anytime soon—not a mobile one, anyway.

That sober reality hasn't deterred other scientists from hatching even more outlandish schemes. One researcher has suggested linking financial centers in different parts of the world for high-frequency trades by using an encoded neutrino beam that takes a shortcut through the Earth. This would give traders a time advantage of up to tens of milliseconds over other means of communication. John Learned and his collaborators Sandip Pakvasa of the University of Hawaii and Anthony Zee of the University of California, Santa Barbara, have proposed using neutrino beams to signal aliens. Since neutrinos travel freely through space for the most part, proponents argue that they would make great messengers between advanced civilizations across the galaxy.

As riveting as some of these potential uses may seem, it is the promise of physics that excites most neutrino hunters. The strongest evidence for something new on the horizon comes from neutrinos having a nonzero mass, contrary to what the standard model presumed. As Fermilab's Boris Kayser says, "All physics is wrong, except where it isn't." What he means is that theories of physics are only valid in the realm where they are meant to apply. "The standard model works very well in the energy domain where it is applicable, but it may not be valid at much higher energies," he points out. The situation is similar to the relationship between Newton's and Einstein's theories of gravity. As Kayser explains, "Newtonian gravity works fine for sending a spacecraft to the Moon, but not if you want to send a probe to the other end of the galaxy and accelerate it to ninety percent of the speed of light. Then you have to use Einstein's version." Similarly, we may need a grander successor to the standard model to make sense of what happens under the most extreme conditions, such as those that prevailed right after the big bang.

In addition to neutrinos turning out to have mass, there are other inklings of anomalies. Data from the Liquid Scintillator Neutrino Detector (LSND) experiment, operated at Los Alamos in the 1990s to probe the chameleonlike nature of neutrinos, suggested they come in four flavors rather than the standard three. The Mini Booster Neutrino Experiment (MiniBooNE) at Fermilab also reported seeing whiffs of a fourth neutrino type. If they exist, neutrinos of the fourth flavor would be even more elusive than the other three types. Heavier than the others, they would be immune to the weak nuclear force, so we would not be able to detect them directly. Such "sterile" neutrinos would still affect their surroundings through gravity, though.

Findings of NASA's Wilkinson Microwave Anisotropy Probe (WMAP), a space observatory mapping the tiny ripples in the afterglow of the big bang, have cast doubt on the existence of a fourth neutrino type, however. The pattern of fluctuations in the cosmic microwave background holds clues to the stew of particles that existed in the early universe. Cosmologists who analyzed the full nine years of WMAP data concluded that there were most likely only three neutrino families at that time. In March 2013, scientists released maps of the cosmic microwave background that are even more exquisite, made by the European Space Agency's Planck spacecraft. Again, they did not find evidence for sterile neutrinos, disappointing some researchers who had hoped for a more exciting result.

However, Janet Conrad of MIT, who was involved in the MiniBooNE experiment, isn't quite ready to give up on the possibility of a fourth neutrino type yet. She contends, "The universe is complicated, and so sterile neutrinos can hide from the Planck results." She says that other effects, such as a large neutrino-antineutrino asymmetry in the early universe, could mask the presence of sterile neutrinos. Besides, cosmologists can't measure the number of neutrino types directly, but have to infer it from a model with many parameters. If they adjust one of those parameters, for instance by increasing the observed expansion rate of the universe slightly, they could accommodate four neutrino types. As Conrad explains, "This is why physicists like to see particles directly" rather than rely on limits from cosmology. "One thing I am sure about: if the direct search [for the sterile neutrino] pans out, the cosmological models will be tweaked to accommodate it," she predicts.

Indeed, neutrino hunting could bring about surprising new revelations about the universe we inhabit, and could do so at a fraction of the cost of building powerful particle accelerators. As Boris Kayser of Fermilab emphasizes, in fact, "Neutrino physics is not terribly expensive. It's more modest in cost than the LHC [Large Hadron Collider]." For the coming decade, physicists based in the United States expect to shift their focus to the so-called intensity frontier, which involves using intense beams of particles and highly sensitive detectors, instead of the "energy frontier," which means building bigger and bigger particle accelerators to reach higher and higher energies. Neutrino physics is the centerpiece of the plans of American physicists. Steven Weinberg concurs that the shift makes sense. "Neutrinos allow us to explore a very interesting frontier at a much lower cost," he explains. Hitoshi Murayama expresses a similar assessment: "High-energy collider experiments in the United States are struggling. The LHC has taken the lead. So Fermilab is betting its future on neutrino experiments, in part because they are cheaper than next-generation colliders."

Unfortunately for the American neutrino hunters, the plans for their next big toy, called the Long-Baseline Neutrino

Experiment, have hit a roadblock. They had expected the National Science Foundation and the Department of Energy (DOE) to share the nearly $2 billion cost of building it. But both institutions have declined to proceed with the original plan, because the price tag was deemed too high at a time of tight budgets. "That's disappointing, but not shocking," says Kayser. "When the NSF pulled out, the DOE couldn't afford it alone. But the DOE is keen to find some way to make it happen." In fact, in December 2012 the DOE gave preliminary approval to build a bare-bones version of the experiment, at half the original price tag, with a smaller detector that sits on the surface instead of underground.

Meanwhile, researchers in other countries are not sitting idle. The Sudbury Neutrino Laboratory in Canada has undergone a major upgrade and is ready to host several experiments. Now called SNOLAB, this underground facility is four times larger than it used to be. In Japan, there are plans to build Hyper-Kamiokande, with a detector ten times bigger than that used for Super-Kamiokande. Europeans are working on new experiments at the Gran Sasso laboratory in Italy and also conducting a design study for a gigantic underground detector called LAGUNA, which may be deployed in a mine in Finland or elsewhere on the continent.

Neutrino hunting has entered a brave new era. Scientists are clamoring for the tools to build on the momentum of the past fifteen years. "Facilities with new capabilities almost always lead to unexpected discoveries," says Kayser. To support his case, he points to Kamiokande, which was initially designed to find out whether protons are unstable over timescales much, much longer than the age of the universe. It did not find any evidence of proton decay, but it did confirm the solar neutrino deficit, register the first neutrinos from an astronomical source other than the Sun—namely, Supernova 1987A—and play a significant role in the discovery of neutrino oscillations. "Neutrino physics could turn out to be even richer than we think," ventures Georg Raffelt.

First invoked by Wolfgang Pauli eighty years ago to dodge a crisis in quantum theory, these pathologically shy particles may herald the exciting next chapter of modern physics, not to mention their burgeoning applications in geology and nuclear monitoring. The findings of today's neutrino hunters have the potential to shake up much of what we know about the universe on its smallest and grandest scales, and quite possibly to upend our theories of particle physics and cosmology. There is good reason for looking to neutrinos for clues to the way forward. MIT's Lindley Winslow put it best: "Whenever anything cool happens in the universe, neutrinos are usually involved."

In fact, neutrinos have baffled and surprised theorists, frustrated and rewarded experimentalists, and fascinated all of us with their strange and fugitive nature. Still, we have come a long way since Pauli reluctantly proposed a "desperate remedy" and Enrico Fermi gave neutrinos their endearing name. Our understanding of neutrinos progressed rather slowly at first. It took nearly a quarter century after Pauli's proposal for Fred Reines and Clyde Cowan to trap the poltergeist for the first time. Ray Davis and John Bahcall puzzled over the solar neutrino deficit for three decades. Despite Bruno Pontecorvo's brilliant insights early on, physicists weren't able to confirm the chameleonlike flavor changes of neutrinos until the end of the twentieth century. But the pace of discovery has picked up in recent years. Physicists have seen the first definitive indications of physics beyond the standard model in the flavor oscillations of neutrinos. Theorists using sophisticated computer simulations have begun to unravel the critical role of neutrinos in the violent deaths of massive stars, while experimentalists stand ready with sensitive detectors for the next supernova in our galactic neighborhood. Researchers are using neutrinos to probe the interior of the Earth and developing neutrino-detection technologies for nuclear monitoring. And, finally, we are on the verge of testing Ettore Majorana's suggestion that neutrinos behave the same way as their antimatter twins, thereby opening the door to solving the great mystery of how matter came to dominate the universe. Neutrino hunters are about to take center stage for the dramatic next chapter of their epic adventure.

TIME LINE

1896: Henri Becquerel discovered radioactivity.

1897–1908: Experiments by Ernest Rutherford, the Curies, Paul Villard, Walter Kaufmann, and Hans Geiger revealed that radioactivity produced three types of emissions: alpha particles (equivalent to helium nuclei), beta particles (electrons), and gamma rays (a highly energetic form of electromagnetic radiation).

1898: Marie and Pierre Curie identified radium and polonium and showed that radioactivity is not limited to uranium.

circa 1908–1927: Otto Hahn, Lise Meitner, James Chadwick, and others found that radioactive beta decays released electrons with a range of energies, raising questions about the law of energy conservation.

1930: Wolfgang Pauli invoked a new neutral particle as "a desperate remedy" to account for the missing energy in beta decay.

1932: James Chadwick discovered the neutron, and Enrico Fermi coined the name "neutrino" to distinguish Pauli's hypothesized particle from the neutron. Carl Anderson discovered the positron in cosmic rays, confirming Paul Dirac's prediction of the existence of antimatter.

1933: Fermi formulated a theory of beta decay that incorporated the neutrino and foreshadowed the weak force.

1937: Ettore Majorana proposed that the neutrino could be its own antiparticle.

1939: Hans Bethe published his seminal paper on energy production in stars but failed to mention neutrinos.

1946: Bruno Pontecorvo proposed that neutrinos produced in nuclear reactors and in the Sun could be detected with chlorine-based experiments.

1955–56: Researchers using the Bevatron in California identified the antiproton and the antineutron.

1956: Frederick Reines and Clyde Cowan definitively detected (anti)neutrinos using a nuclear reactor as the source.

1956–57: T. D. Lee and C. N. Yang suggested that weak interactions might not conserve parity. C. S. Wu and her colleagues found evidence of parity violation in beta decay.

1957: Bruno Pontecorvo suggested that neutrinos might not necessarily be massless and hypothesized about neutrino oscillations for the first time.

1962: Leon Lederman, Melvin Schwartz, and Jack Steinberger detected the muon neutrino.

1964: James Cronin, Val Fitch, and their colleagues observed CP violation in neutral particles called kaons.

1964: John Bahcall and Ray Davis suggested the feasibility of measuring neutrinos from the Sun and made the case for the Homestake mine experiment.

1968: Ray Davis reported that he had detected solar neutrinos, but only about a third as many as the solar models predicted, thus giving rise to the "solar neutrino problem," which remained unsolved for three decades.

1969: Vladimir Gribov and Bruno Pontecorvo proposed that

neutrino oscillations between flavors could account for the missing solar neutrinos.

circa 1974: The standard model of particle physics emerged.

1976: Martin Perl and colleagues discovered the tau particle, leading to the suggestion that a third neutrino flavor associated with the tau exists.

1978–86: Theorists Lincoln Wolfenstein, Stanislav Mikheyev, and Alexei Smirnov found that the presence of matter could enhance neutrino oscillations.

1987: Kamiokande and two other experiments recorded a total of two dozen neutrinos from Supernova 1987A in the Large Magellanic Cloud, a satellite galaxy of the Milky Way located 160,000 light-years away. These were the first neutrinos to be detected from beyond the solar system.

1989: Kamiokande provided independent confirmation of the solar neutrino deficit.

1995: Frederick Reines was awarded a share of the Nobel Prize for his detection of neutrinos with Clyde Cowan.

1998: Super-Kamiokande reported strong evidence of oscillations in atmospheric neutrinos. The oscillations implied that neutrinos have nonzero mass, thus suggesting physics beyond the standard model.

2000: Physicists at Fermilab made direct observations of tau neutrinos for the first time.

2001–2002: The Sudbury Neutrino Observatory reported compelling evidence that neutrino oscillations are responsible for the solar neutrino deficit. Data from the KamLAND experiment, which measured antineutrinos from nuclear reactors, provided independent confirmation of neutrino oscillations.

2002: Ray Davis and Masatoshi Koshiba won shares of the

Nobel Prize for their roles in the detection of neutrinos from the Sun and Supernova 1987A.

2005: KamLAND researchers reported measuring "geoneutrinos" produced by radioactive elements in the Earth's interior.

2011–2012: Tokai-to-Kamioka (T2K), Double Chooz, and RENO collaborations presented evidence that the third mixing angle (θ_{13}) is nonzero, and the Daya Bay experiment measured its value.

2012: Two experiments at the Large Hadron Collider at CERN discovered the long-sought Higgs boson, confirming a key prediction of the standard model.

2013: Planck spacecraft's observations of the cosmic microwave background favored the existence of only three flavors of light neutrinos, and provided a new limit on the sum total of the three neutrino masses when combined with other cosmological data.

GLOSSARY

alpha ray (or alpha particle): A bundle comprising two protons and two neutrons; the same as the nucleus of helium. When a radioactive nucleus releases an alpha particle, the nucleus transforms into a different element, which comes two places earlier in the periodic table.

annihilation: The process that occurs when a particle meets its antiparticle. For example, when an electron and a positron annihilate, they produce gamma rays.

antimatter: Material composed of antiparticles, which have the same mass as their ordinary-matter counterparts, but the opposite charge and spin.

artificial radioactivity: Radioactivity induced in an otherwise stable material by bombarding it with high-speed particles.

atmospheric neutrinos: Neutrinos produced when cosmic rays hit the upper atmosphere of the Earth.

baryons: A family of subatomic particles, each of which is composed of three quarks. Baryons are subject to the strong force.

beta decay: The spontaneous transformation of an atomic nucleus that converts a neutron into a proton while releasing an electron and an antineutrino.

beta rays: Fast-moving electrons or positrons emitted by some radioactive nuclei.

Betelgeuse: A red supergiant star that marks the hunter's right shoulder in the Orion constellation; it is a prime candidate to explode as a supernova someday.

big bang: The explosive birth of the cosmos. Evidence for the big bang theory includes the observed expansion of the universe, the cosmic microwave background, and the abundance ratios of the light elements.

black hole: A region of space from which nothing, not even light, can escape. A black hole could form when a massive star's core contracts drastically after a supernova explosion.

blue supergiant: A late stage in evolved massive stars, characterized by high surface temperatures (tens of thousands of degrees Celsius) and large radii (tens of times that of the Sun).

CERN: The acronym for the European Organization for Nuclear Research, from the French name for the provisional council that set up the laboratory, *Conseil Européen pour la Recherche Nucléaire*.

chain reaction: Series of (nuclear) reactions, with one triggering the next. Examples include the p-p chain and the CNO cycle in stars. Chain reactions also make atomic bombs and nuclear reactors possible.

Cherenkov radiation: The characteristic pale-blue emission of particles such as electrons moving through water or ice at speeds greater than that of light in the same medium (though still slower than the speed of light in a vacuum).

cloud chamber: An apparatus that detects charged particles by the trails they leave in supersaturated water or alcohol vapor. Carl Anderson used cloud chambers to discover the positron and the muon in the 1930s.

CNO cycle: A set of fusion reactions using carbon, nitrogen, and oxygen as catalysts to convert hydrogen into helium. The CNO cycle is the dominant energy production mechanism in stars heftier than about 1.3 solar masses.

cosmic microwave background (CMB): The afterglow of the big bang; radiation left over from the very hot early epoch about 380,000 years after the birth of the universe. The CMB was discovered in 1965, and has been mapped in exquisite detail since.

cosmic neutrino background: The sea of low-energy "relic" neutrinos that is thought to permeate the universe. These neutrinos were produced in the big bang and decoupled from ordinary matter when the universe was barely a couple of seconds old.

cosmic rays: Fast-moving particles, either protons or atomic nuclei, from deep space that bombard the Earth from all directions. Growing evidence suggests that supernova remnants are the primary source of cosmic rays.

cosmology: The study of the origin, evolution, large-scale structure, and eventual fate of the universe.

CP violation: A violation of the CP (or charge-parity) symmetry, which postulates that the laws of physics should be the same if you swap a particle with its antiparticle and left with right. Understanding CP violation is central to figuring out how matter came to predominate over antimatter in the early universe.

Crab Nebula: The gaseous remnant of the supernova that was seen in 1054, located in the constellation Taurus, with a neutron star near its center.

dark matter: Invisible material whose existence is inferred from its gravitational influence on visible matter in galaxies and galaxy clusters. Dark matter appears to be much more abundant than normal matter, but what it's made of remains a mystery. At one time, scientists thought that neutrinos could be a major constituent, but now we know that their mass is too small.

deuterium: An isotope of hydrogen whose nucleus contains a neutron as well as a proton, instead of a lone proton.

double beta decay: The rare process in which two neutrons in the same nucleus decay into two protons simultaneously, releasing two electrons and two antineutrinos. If neutrinos are their own antiparticles, as Ettore Majorana suggested, the antineutrino emitted by one neutron can be absorbed immediately by the other, resulting in what's called "neutrinoless double beta decay." If physicists confirm that neutrinoless double beta decay does indeed occur in nature, that would prove Majorana right and open the door to explaining the matter-antimatter asymmetry.

electromagnetic radiation: Radiation associated with oscillating electric and magnetic fields, including radio waves, microwaves, infrared light, visible light, ultraviolet light, X-rays, and gamma rays.

electron: A negatively charged elementary particle that is usually in orbit around the nucleus of an atom. It belongs to the lepton family.

element: A fundamental chemical substance consisting of one type of atom and characterized by the number of protons in its nucleus. Of the ninety-eight naturally occurring elements, the lightest—hydrogen, helium, and traces of lithium and beryllium—were created within minutes of the big bang, while all others were produced in stars and supernovae.

energy conservation: The principle (or law) in physics that states that energy can neither be created nor destroyed, although it can change form and convert into mass and back according to the relation $E = mc^2$.

Eta Carinae: A very massive star (with a likely companion) that may explode as a supernova in the future. It has shed outer layers in past eruptions and sits at the center of a gigantic nebula.

fission: The breaking up of an atomic nucleus into lighter pieces, often releasing a large amount of energy and particles such as neutrons.

flavor: The whimsical name given to different varieties of a particle. There are three flavors of neutrinos associated with the electron, the muon, and the tau particle.

fusion: The joining together of light atomic nuclei to form a heavier nucleus, accompanied by the release of energy (and possibly other particles). Fusion powers most stars.

galaxy: A large collection of stars, typically hundreds of millions to hundreds of billions, as well as dust and gas, held together by gravity. See also Milky Way.

gamma rays: Radiation even more energetic than X-rays, with wavelengths shorter than those of X-rays. Some radioactive materials emit gamma rays. Gamma rays are also produced when matter and antimatter come together and annihilate each other.

Geiger counter: A simple instrument that detects radiation, such as alpha or beta particles or gamma rays emitted by radioactive materials.

general theory of relativity: A theory that provides a unified geometric description of gravity, space, and time, first proposed by Albert Einstein in 1915. Over the years, scientists have tested many of its predictions, such as bending of light by gravity and gravitational time delay.

gravitational waves: Ripples in the fabric of space-time that propagate as waves. Mergers of black holes or neutron stars would generate strong gravitational waves. Predicted by Einstein's general theory of relativity, they have not been observed directly yet, but there is strong indirect evidence for their existence.

geoneutrinos: Neutrinos produced by radioactive elements inside the Earth.

heavy water: Chemically the same as regular water (H_2O), but with the two hydrogen atoms replaced by deuterium (hence the formula D_2O).

helioseismology: The study of waves in the Sun in order to learn about its interior.

helium: The second lightest and second most common element in the universe, with two protons in its nucleus. Stars fuse hydrogen into helium; helium itself fuses into carbon and oxygen.

Higgs boson: The particle associated with the Higgs field in the standard model that is responsible for endowing some particles with mass. In the summer of 2012, two experiments at the Large Hadron Collider reported evidence of its existence.

hydrogen: The lightest and most abundant element in the universe. The most common form of hydrogen contains only a single proton in its nucleus. The rare form, called "heavy hydrogen" or deuterium, contains a neutron in addition to the proton.

hydrophone: Microphone designed for use underwater.

isotopes: Different types of atoms of the same chemical element, each harboring a different number of neutrons. Some isotopes are radioactive, and thus decay into other types of atoms by spontaneously emitting particles and radiation. Deuterium is an isotope of hydrogen.

Kamiokande: Stands for the Kamioka *n*ucleon *d*ecay experiment. Originally built to look for the possible decay of the proton over very long timescales, this experiment—and its upgraded version, dubbed Super-Kamiokande—helped advance our understanding of neutrinos, notably through the detection of neutrinos from Supernova 1987A and the measurement of neutrino oscillations.

Large Hadron Collider: The world's most powerful particle accelerator, located at the CERN laboratory near Geneva.

leptons: A family of elementary particles, including the electron and the neutrino, which are not subject to the strong force but interact via the weak force.

light-year: The distance that light travels in a year, just under 10 trillion kilometers (or about 6 trillion miles).

Magellanic Clouds: The Large and Small Magellanic Clouds (LMC and SMC) are two irregularly shaped satellite galaxies of the Milky Way. Supernova 1987A occurred in the LMC.

Manhattan Project: The research and development program that produced the first atomic bombs during the Second World War. Many top physicists worked on the project under the auspices of the U.S. government, with additional support from the United Kingdom and Canada.

mantle: The layer that is located between the crust and the core of the Earth (or other terrestrial planets).

mass: The amount of matter in an object.

matter effect: See MSW effect.

microsecond: One-millionth of a second.

Milky Way: Our galaxy, which consists of a flattened disk of stars, dust, and gas with spiral arms, a central bulge, and a large spherical halo. The Sun is located in the outskirts of the galaxy's disk, and there is likely to be a massive black hole at its center.

MSW effect (or matter effect): Named after its discoverers Stanislav Mikheyev, Alexei Smirnov, and Lincoln Wolfenstein, the process by which the presence of matter enhances neutrino oscillations.

muon: A short-lived, negatively charged particle with a mass of about two hundred times that of the electron. Together with the electron, the tau particle, and the neutrino, it belongs to the lepton family.

neutrino oscillations: A quantum mechanical phenomenon by which neutrinos morph between flavors. The existence of oscillations implies that neutrinos have nonzero mass.

neutron: A subatomic particle with no electric charge and a mass slightly larger than that of a proton, usually found in atomic nuclei.

neutron star: Dense, compact remnant of a massive star that exploded as a supernova. Made almost entirely of neutrons.

new physics: A term used to describe phenomena that the standard model does not account for.

Noether's theorem: In simple terms, mathematician Emmy Noether's proposition that a symmetry in nature implies a law of conservation, and vice versa.

nucleon: A particle in the atomic nucleus, i.e., either a proton or a neutron.

nucleus (plural: nuclei): The dense central core of an atom, consisting of protons and neutrons, bound together by the strong force.

pair production: The creation of a particle and its antiparticle from energy (as permitted by the relation $E = mc^2$, where E stands for energy, m for mass, and c for the speed of light in a vacuum).

particle accelerator: A device used to accelerate particles (usually in narrow beams) to very high speeds.

Pauli exclusion principle: The rule, formulated by Wolfgang

Pauli, that no two fermions (particles with half-integer spin) can occupy the same quantum state. It explains how white dwarfs and neutron stars hold up against gravity without shrinking further.

photon: A quantum of light (or other electromagnetic radiation).

phototube: A device sensitive to light. Used for registering Cherenkov radiation emitted when a neutrino creates an electron or a muon in water or ice, for example.

pitchblende: A dark-colored, radioactive, uranium-rich mineral.

plate tectonics: Slow, large-scale motions of plates of the Earth's (or another rocky planet's) crust.

positron: A positively charged particle with the same mass as the negatively charged electron; the antielectron.

p-p chain (or proton-proton chain): A series of fusion reactions that converts hydrogen into helium inside the Sun (and other low-mass stars).

proton: A positively charged subatomic particle found in the nucleus of every atom. The number of protons determines each element: e.g., hydrogen atoms have one proton, helium atoms have two protons, carbon atoms have six protons.

quantum (plural: quanta): A discrete bundle of energy that an atom can absorb or emit. A photon is a single quantum of light.

quantum mechanics: The theory of physics, developed in the early part of the twentieth century, that deals with phenomena on microscopic scales, and describes the interactions between matter and radiation in terms of quanta and probabilities.

quark: One of a class of elementary particles that combine to make protons and neutrons. Quarks come in six flavors.

radioactivity: A process by which an unstable atomic nucleus transforms into a different state by releasing particles and/or energy. There are several types of radioactive decay; see alpha ray, beta decay, gamma rays.

radiometric dating: A technique used to date rocks and meteorites by comparing the abundances of radioactive isotopes and their decay products and using known decay rates. Scientists have relied on radiometric dating to determine the age of the Earth and the solar system.

red supergiant: The bloated, helium-burning phase of stars more massive than about ten solar masses, characterized by enormous radii (hundreds of times that of the Sun) and relatively cool surface temperatures (a few thousand degrees Celsius).

Sanduleak −69° 202: The catalog name of the blue supergiant star that blew up as Supernova 1987A.

Schrödinger's cat: A thought experiment devised by Erwin Schrödinger to illustrate the strangeness of quantum mechanics, in which a cat in a box could be both alive and dead simultaneously until an observer intervenes by looking inside.

seesaw mechanism: A theoretical model proposed to understand the masses of neutrinos relative to other elementary particles; it predicts the existence of heavy counterparts to the familiar light neutrinos.

solar model: A physical model describing the solar interior and the nuclear processes at the Sun's core.

solar neutrino problem: The long-standing discrepancy between the number of solar neutrinos measured and the predictions of theoretical calculations. The mismatch was finally resolved around the year 2000, thanks to the discovery of neutrino oscillations.

solar neutrinos: Neutrinos produced by nuclear reactions inside the Sun.

special theory of relativity: The theory, proposed by Albert Einstein in 1905 building on earlier ideas of Galileo and others, that describes objects moving at constant speeds relative to each other. It posits that the passage of time is relative while the speed of light is not.

standard model: The theoretical framework of physics that describes the material world in terms of a limited set of fundamental particles, their antimatter twins, and "force carriers" such as photons. Developed over several decades and formalized in the 1970s, its predictions have been verified by numerous experiments. However, it does not account for neutrinos having mass or for other puzzling phenomena such as "dark matter."

strong force: One of the four fundamental forces of nature; it operates over very small distances. It holds quarks together to make neutrons and protons, and in turn binds neutrons and protons into atomic nuclei.

Super-Kamiokande (or Super-K): See Kamiokande.

superluminal: Faster than the speed of light.

supernova: A huge explosion of a massive star at the end of its life or of a stellar cinder that accumulates material from a companion. Elements heavier than iron (e.g., gold) are produced only in supernovae.

Supernova 1987A: The supernova in the Large Magellanic Cloud seen in the year 1987. Detectors on Earth also recorded neutrinos emitted during this supernova explosion.

theory: A hypothesis that has withstood experimental and/or observational tests.

time dilation: In the theory of relativity, the difference in elapsed time measured by two observers, either because one is traveling at a high speed relative to the other or because the two experience different gravitational potentials.

tritium: A rare isotope of hydrogen whose nucleus contains two neutrons and a proton, instead of a lone proton, as is the case with ordinary hydrogen.

uncertainty principle: The mathematical relation, sometimes called the Heisenberg uncertainty principle, which states that there is a fundamental limit to how precisely we can measure the properties of a particle (or a system of particles). For example, the more precisely we pin down a particle's position, the less well we can know its momentum, and vice versa.

variable star: A star whose brightness changes with time, perhaps because it swells and shrinks in size, or because a companion eclipses it periodically.

wavelength: The distance between two successive crests or troughs of a wave.

wave-particle duality: The concept in quantum mechanics that particles exhibit both wave and particle properties. Light can be thought of as consisting simultaneously of both waves and particles.

weak force: One of the four fundamental forces of nature; it governs radioactivity.

white dwarf: The compact, dense core of a star that has shed its outer layers. It is the end state of stars comparable in mass to our Sun, which are not massive enough to explode as supernovae.

X-rays: Radiation somewhat less energetic than gamma rays, with wavelengths shorter than those of ultraviolet light but longer than those of gamma rays.

NOTES

Two books on the subject—*Spaceship Neutrino* by Christine Sutton (Cambridge University Press, 1992) and *Neutrino* by Frank Close (Oxford University Press, 2010)—provide excellent overviews, with good coverage of the early developments and the solar neutrino problem, respectively. Isaac Asimov's *The Neutrino: Ghost Particle of the Atom* (New York: Doubleday, 1966) is of historical interest.

1. THE HUNT HEATS UP

3 *Jens Stoltenberg*: Video of the Norwegian prime minister's speech on the centenary of Amundsen's arrival at the South Pole is available at www .telegraph.co.uk/news/worldnews/antarctica/8956091/100th-anniversary -of-Roald-Amundsen-reaching-South-Pole-is-honoured.html. *The Antarctic Sun* reports on the celebration are at http://antarcticsun.usap .gov/features/contentHandler.cfm?id=2555 and http://antarcticsun.usap .gov/features/contenthandler.cfm?id=2554.

3 *centenary of Amundsen*: There are many excellent books on Antarctic exploration, and *The New York Times* published an article titled "Amazing Race to the Bottom of the World" by John Noble Wilford on December 12, 2011, to mark the centenary.

4 *I went to Antarctica*: See my article "The Meteorite Hunters" in the November/December 2011 issue of *Muse* magazine (Chicago: Carus Publishing Company; www.musemagkids.com).

6 *small flags that*: Francis Halzen kindly sent me photographs taken by his colleagues so that I could see what IceCube looked like on the Amundsen centennial.

6 *IceCube*: Description of IceCube is based, in part, on a telephone interview with Francis Halzen conducted by the author on December 12, 2011, and on material on the project website at http://icecube.wisc.edu/.

6 *phototubes*: Though I have used the term "phototube" for simplicity, in fact these are photomultiplier tubes (PMTs for short); incident light generates an electric current in the PMT, which is then amplified up to tens of millions of times to make the detection easier.

8 *"If you're trying"*: This Janet Conrad quote is from a telephone interview conducted by the author on March 4, 2013.

8 *Boris Kayser*: Quotes are from a telephone interview conducted by the author on August 9, 2012.

9 *Hitoshi Murayama*: Quotes are from a Skype interview with the author on March 28, 2012.

10 *Klaatu*: Lyrics of their song "Little Neutrino" are available at www.klaatu.org/lyrics/347est_lyrics.html.

10 *popular sitcom*: Quotes are from the fourth episode, titled "The Griffin Equivalency," of the second season of *The Big Bang Theory*.

11 *OPERA*: The initial CERN press release and the subsequent updates are available at http://press.web.cern.ch/press-releases/2011/09/opera-experiment-reports-anomaly-flight-time-neutrinos-cern-gran-sasso.

11 *"If the Europeans"*: Quoted from Michael D. Lemonick, "Was Einstein Wrong? A Faster-Than-Light Neutrino Could Be Saying Yes," *Time*, September 23, 2011, www.time.com/time/health/article/0,8599,2094665,00.html.

11 *"If true, it is"*: Quoted from Dennis Overbye, "Tiny Neutrinos May Have Broken Cosmic Speed Limit," *The New York Times*, September 23, 2011.

12 *questioning whether* $E = mc^2$: Video of the song by Corrigan Brothers and Pete Creighton is at www.youtube.com/watch?v=vpMY84T8WY0.

12 *Andrew Cohen*: See "New Constraints on Neutrino Velocities" by Andrew Cohen and Sheldon Glashow at http://arxiv.org/abs/1109.6562.

13 *Most media reports*: See, e.g., Gautam Naik, "Unreal Finding May Be Just That," *The Wall Street Journal*, February 24, 2012, and Dennis Overbye, "The Trouble with Data That Outpaces Theory," *The New York Times*, March 26, 2012.

16 *"We're right on"*: Kate Scholberg quotes are from a telephone interview conducted by the author on March 16, 2012.

16 *Francis Halzen*: Quotes and biographical information are from a tele-

phone interview conducted by the author on December 12, 2011. Description of the early days of the AMANDA experiment and the quotes "Learned immediately appreciated," "To have your career," and "a nearly meaningless blur" are from Halzen's essay "Antarctic Dreams," first published in *The Sciences* (March–April 1999): 19–24. For additional background on Halzen's scientific interests and the development of AMANDA, see Halzen's essay "Ice Fishing for Neutrinos" at http://icecube.berkeley.edu/amanda/ice-fishing.html.

17 *John Learned*: Biographical information and quotes are from a Skype interview conducted by the author on March 6, 2013.

22 *"PeV events"*: See Aya Ishihara, "Ultra-High Energy Neutrinos with Ice-Cube," *Nuclear Physics B Proceedings Supplement* (2012), available at www.ppl.phys.chiba-u.jp/research/IceCube/ThePeVNeutrinoDetection/IceCubeEHE2012_v6.pdf; and "High-energy (PeV) neutrinos observed!," blog entry by Spencer Klein, *Neutrino Hunting in Antarctica*, August 8, 2012, http://antarcticaneutrinos.blogspot.ca/2012/08/high-energy-pev-neutrinos-observed.html. Additional information and quotes come from e-mail interviews with Francis Halzen (March 18–19, 2013) and Spencer Klein (March 18–19, 2013).

23 *gamma ray bursts*: For a good overview, see Joshua S. Bloom, *What Are Gamma Ray Bursts?* Princeton Frontiers in Physics (Princeton, NJ: Princeton University Press, 2011).

2. A TERRIBLE THING

27 *special and general theories of relativity*: There are many good popular books on relativity. For example, Russell Stannard, *Relativity: A Very Short Introduction* (New York: Oxford University Press, 2008) covers the basics well.

28 *"time dilation"*: Joseph Hafele and Richard Keating conducted one of the more famous experiments of relativistic time dilation in 1971 by flying four atomic clocks aboard commercial flights. They reported their findings in a pair of papers in *Science*: J. C. Hafele and R. E. Keating, "Around-the-World Atomic Clocks: Predicted Relativistic Time Gains" and "Around-the-World Atomic Clocks: Observed Relativistic Time Gains," *Science* 177 (July 14, 1972): 166–68 and 168–70.

28 *quantum mechanics*: I remember reading *In Search of Schrödinger's Cat: Quantum Physics and Reality* by John Gribbin (New York: Bantam

Books, 1984) as a teenager. The book is a lively introduction to the weird world of quantum physics.

30 *Wolfgang Pauli*: The primary source of biographical information and quotes on and from Pauli is Charles P. Enz, *No Time to Be Brief: A Scientific Biography of Wolfgang Pauli* (New York: Oxford University Press, 2002).

31 *"I have around me,"* . . . *"Whoever studies this"*: As quoted in Arthur I. Miller, *Deciphering the Cosmic Number: The Strange Friendship of Wolfgang Pauli and Carl Jung* (New York: Norton, 2009).

35 *"If it had been a bullfighter"*: As quoted in Enz, *No Time to Be Brief.*

36 *Ernest Rutherford*: There is a brief biographical sketch online at www.nobelprize.org/nobel_prizes/chemistry/laureates/1908/rutherford-bio.html, and his 1908 Nobel Prize Lecture, "The Chemical Nature of Alpha Particles from Radioactive Substances," is available online at www.nobelprize.org/nobel_prizes/chemistry/laureates/1908/rutherford-lecture.html.

37–38 *"It would, nevertheless"* . . . *"My husband and I"* . . . *"One of our joys"*: Quoted in Marie Curie, *Pierre Curie, with Autobiographical Notes* (New York: Macmillan, 1923); online at http://etext.virginia.edu/toc/modeng/public/CurPier.html.

39 *"We may say"*: Niels Bohr delivered the Faraday Lecture to the Fellows of the Chemical Society in London on May 8, 1930, and it was published as "Chemistry and the Quantum Theory of Atomic Constitution" in the *Journal of the Chemical Society* (1932): 349–84.

40 *"Do you intend"*: As quoted in Gino Segrè, *Faust in Copenhagen: A Struggle for the Soul of Physics* (New York: Viking, 2007), 194.

40 *"What if someone"*: As quoted in Miller, *Deciphering the Cosmic Number.*

40 *"Dear Radioactive Ladies and Gentlemen"*: The original German text of the letter is available at the Pauli Archive at CERN and online at http://cds.cern.ch/record/83282/files/meitner_0393.pdf.

42 *"I have done a terrible thing"*: As far as I could find, this quote attributed to Pauli was first reported by Fred Hoyle, who heard it from fellow astronomer Walter Baade. Pauli made the remark to his close friend Baade when the latter stayed with Pauli in Hamburg. Hoyle's telling of the story is published in *Proceedings of the Royal Society of London A* 301 (October 17, 1967): 171.

42 *"The matter still"*: From Wolfgang Pauli, "On the Earlier and More Recent History of the Neutrino," trans. Gabriele Zacek (1957), in *Neutrino*

Physics, 2nd ed., ed. Klaus Winter (Cambridge: Cambridge University Press, 2000), 1–21.

42 *"A new inhabitant"*: From a report on the American Association for the Advancement of Science meeting written by a staff correspondent of *The New York Times* and published in the June 17, 1931, issue of the newspaper.

42–43 *"I recently fell"* . . . *"This, I am afraid"*: From Enz, *No Time to Be Brief*.

43 *Enrico Fermi*: The primary source of biographical information on Enrico Fermi is the biography written by his wife, Laura Fermi, *Atoms in the Family: My Life with Enrico Fermi* (Chicago: University of Chicago Press, 1954). Also useful is the biographical memoir of Fermi by Samuel K. Allison (Washington, D.C.: National Academy of Sciences, 1957), available online at www.nasonline.org/publications/biographical-mem oirs/memoir-pdfs/fermi-enrico.pdf. On a side note, I happen to have a distant connection to Fermi: he is my "academic great-grandfather." One of my PhD advisers at Harvard, Giovanni Fazio, studied under Robert Schluter at MIT, who in turn was a student of Fermi's at the University of Chicago.

44 *"immediately expressed a lively interest"*: From Pauli, "History of the Neutrino."

45 *"When the hard-boiled rationalist"*: As quoted in Miller, *Deciphering the Cosmic Number*, which charts the fascinating friendship between Pauli and Jung. So does David Lindorff in *Pauli and Jung: The Meeting of Two Great Minds* (Wheaton, IL: Quest Books, 2004).

45 *"that foolish child"*: As quoted in Enz, *No Time to Be Brief*, from a letter Pauli wrote to biophysicist Max Delbrück on October 6, 1958, just two months before Pauli's death, and available in the Pauli Archive at CERN.

46 *"the rest mass of the neutrino"*: Fermi derived an estimate of neutrino mass in his classic paper, "An Attempt at a Theory of Beta Radiation," published in the German journal *Zeitschrift für Physik* 88 (1934): 161.

48 *"the Italian navigator"*: As recounted in Arthur H. Compton, "The Birth of Atomic Power," *Bulletin of the Atomic Scientists* 9, no. 1 (February 1953): 10–12.

49 *"no practically possible way"*: From a letter to the editor of *Nature* by Hans Bethe and Rudolf Peierls, "The 'Neutrino,'" *Nature* 133 (April 1934): 532.

50 *"neutrinos seem to provide"*: Quoted in Helge Kragh, *Dirac: A Scientific Biography* (Cambridge: Cambridge University Press, 1990).

3. GHOST CHASING

53 *"It seems therefore"*: Quoted in Lise Meitner and O. R. Frisch, "Disintegration of Uranium by Neutrons: A New Type of Nuclear Reaction," *Nature* 143 (February 11, 1939): 239–40.

54 *Manhattan Project*: The best popular account of the development of nuclear weapons is Richard Rhodes, *The Making of the Atomic Bomb* (New York: Simon & Schuster, 1987).

55 *Bruno Pontecorvo*: Historian Simone Turchetti's *The Pontecorvo Affair: A Cold War Defection and Nuclear Physics* (Chicago: University of Chicago Press, 2012) recounts Pontecorvo's life and discusses the circumstances surrounding his defection in detail. Other useful sources include Randy Kennedy's *New York Times* obituary (September 28, 1993), and Luisa Bonolis, "Bruno Pontecorvo: From Slow Neutrons to Oscillating Neutrinos," *American Journal of Physics* 73, no. 6 (June 2005): 487–99.

58 *"It occurred to me"*: Quoted in Bruno Pontecorvo, "Fifty Years of Neutrino Physics: A Few Episodes," in *Neutrino Physics and Astrophysics*, ed. Ettore Fiorini (New York: Plenum Press, 1982).

60 *Klaus Fuchs*: See "The Atom Spy Case" on the FBI website at www.fbi.gov/about-us/history/famous-cases/the-atom-spy-case.

60 *Richard Feynman*: See "The Feynman Files" at www.muckrock.com/news/archives/2012/jun/06/feynman-files-professors-invitation-past-iron-curt/.

61 *newspapers at the time*: The commotion and suspicions surrounding Pontecorvo's disappearance were reflected in media reports at the time, including "Atom Man Flies Away" in *Daily Express* (London; October 21, 1950), "Atomic Expert Missing" in *The Manchester Guardian* (October 21, 1950), "Top Atom Expert Flees to Russia" in *The Mail* (Adelaide, Australia; October 21, 1950) and "Atom Scientist Mystery" in *The Sydney Morning Herald* (October 23, 1950).

62 *Even the BBC*: From "On This Day 1950: Hunt for missing atomic scientist," online at http://news.bbc.co.uk/onthisday/hi/dates/stories/october/27/newsid_3091000/3091390.stm.

63 *Pontecorvo revealed that*: As recounted in Charles Richards, "Confessions of an Atom Spy," *The Independent*, August 2, 1992.

64 *Fred Reines*: Biographical details of Fred Reines are based on his own recollections at www.nobelprize.org/nobel_prizes/physics/laureates /1995/reines-autobio.html, and on William Kropp, Jonas Schultz, and Henry Sobel, *Frederick Reines, 1918–1998: A Biographical Memoir* (Washington, D.C.: National Academy of Sciences, 2009).

67 *With this design*: Primary sources for describing the project are Fred Reines's 1995 Nobel Prize Lecture, "The Neutrino: From Poltergeist to Particle" (online at www.nobelprize.org/nobel_prizes/physics/laureates /1995/reines-lecture.html); E. C. Anderson, "The Reines-Cowan Experiments: Detecting the Poltergeist," *Los Alamos Science* 25 (1997); and Robert G. Arns, "Detecting the Neutrino," *Physics in Perspective* 3 (2001): 314–34.

68 *"We would return"*: From Clyde Cowan, *Anatomy of an Experiment: An Account of the Discovery of the Neutrino* (Washington, D.C.: Smithsonian Institution, 1964).

69 *"Certainly your new"*: Fermi's letter to Reines, quoted in Reines's Nobel Lecture.

70 *"Those days at Hanford"*: From Frederick Reines, "Neutrinos to 1960— Personal Recollections," Proceedings of the International Colloquium on the History of Particle Physics, *Journal de Physique* 43, no. C8 (December 1982): 237–60.

70 *made it to the popular press*: As described in Robert G. Arns, "Detecting the Neutrino."

72 *Pauli's message failed*: According to Enz, *No Time to Be Brief*.

4. SUN UNDERGROUND

75 *Ray Davis*: Biographical details, the quotes "It was to please [my mother]" and "To my surprise," and descriptions of his experiments are based in large part on his own recollections at www.nobelprize.org/nobel_prizes /physics/laureates/2002/davis-autobio.html and his 2002 Nobel Lecture, "A Half-Century with Solar Neutrinos," online at www.nobelprize .org/nobel_prizes/physics/laureates/2002/davis-lecture.pdf. Another useful reference is Kenneth Lande, "The Life of Raymond Davis, Jr., and the Beginning of Neutrino Astronomy," *Annual Review of Nuclear and Particle Science* 59 (November 2009): 21–39.

78 *Hans Bethe*: Silvan S. Schweber has published a new biography of Bethe, *Nuclear Forces: The Making of the Physicist Hans Bethe* (Cambridge:

Harvard University Press, 2012). The quotes "mathematics seemed to" and "The best thing" and biographical details about Bethe are from Schweber's article "The Happy Thirties," in *Hans Bethe and His Physics*, ed. Gerald E. Brown and Chang-Hwan Lee (Singapore: World Scientific, 2006).

81 *CNO cycle*: There is a description of the CNO cycle in John Bahcall, "How the Sun Shines" (2000), online at www.nobelprize.org/nobel_prizes /physics/articles/fusion/index.html).

81 *"I did not, contrary to legend"*: From Hans Bethe, "My Life in Astrophysics," in Brown and Lee, *Hans Bethe and His Physics*.

82 *Davis used his experiment*: History of Davis's experiments and collaboration with Bahcall is based on a number of sources, including Ray Davis's Nobel Lecture, "A Half-Century with Solar Neutrinos"; Christine Sutton, *Spaceship Neutrino*; John Bahcall, "Solving the Mystery of the Missing Neutrinos" (2004), online at www.nobelprize.org/nobel_prizes /physics/articles/bahcall/.

82 *"One would not write"*: As quoted in John N. Bahcall and Raymond Davis, Jr., "An Account of the Development of the Solar Neutrino Problem," in *Essays In Nuclear Astrophysics*, ed. Charles A. Barnes, Donald D. Clayton, and David Schramm (Cambridge: Cambridge University Press, 1982).

84 *"The probability of a negative result"*: As quoted by Davis at www.nobel prize.org/nobel_prizes/physics/laureates/2002/davis-lecture.pdf.

84 *John Bahcall*: The quote "It was the hardest" and some biographical details are from Bahcall, "Two Secrets," commencement address to physics and astronomy graduates of the University of California, Berkeley (2001), online at www.sns.ias.edu/~jnb/Papers/Popular/secrets.htm.

86 *"When I got to Caltech"*: From Bahcall's interview with *Nova*, the PBS program (February 21, 2006), online at www.pbs.org/wgbh/nova/physics /solar-neutrinos.html.

87 *"Ray's greatest political"*: From John Bahcall, "Ray Davis: The Scientist and the Man," *Nuclear Physics B (Proc. Suppl.)* 48 (1996): 281–83.

88 *"These tank people" . . . "ten minutes' time"*: As quoted in Bahcall and Davis, "An Account of the Development of the Solar Neutrino Problem," in Barnes, Clayton, and Schramm, *Essays In Nuclear Astrophysics*.

89 *"just plumbing"*: From Bahcall's interview with *Nova*.

89 *"as a nonchemist"*: From John Bahcall, "Neutrinos from the Sun," *Scientific American*, July 1969, 28.

93 *"All the people"*: From Bahcall, "The Scientist and the Man."

93 *"one of the biggest embarrassments"*: From George Johnson, "Elusive

Particles Continue to Puzzle Theorists of the Sun," *The New York Times*, June 9, 1998.

94 *"had the whole auditorium"*: From Bahcall's interview with *Nova*.

94 *Masatoshi Koshiba*: See David DeVorkin's interview with Koshiba on August 30, 1997, Niels Bohr Library & Archives, Center for the History of Physics, American Institute of Physics; edited transcript online at www.aip.org/history/ohilist/24870.html.

5. COSMIC CHAMELEONS

97 *Three physicists*: Leon Lederman, Melvin Schwartz, and Jack Steinberger received the 1988 physics Nobel Prize for the discovery of the muon neutrino: www.nobelprize.org/nobel_prizes/physics/laureates /1988/.

97 *tau particle*: Fermilab press release, July 20, 2000 announcing the discovery of the tau neutrino: www.fnal.gov/pub/presspass/press_releases /donut.html.

98 *neutrinos oscillating*: Gribov and Pontecorvo published their paper, "Neutrino Astronomy and Lepton Charge," in *Physics Letters B* 28, no. 7 (1969): 493–96.

100 *MSW effect*: Also known as the "matter effect."

100 *"The MSW effect is a beautiful idea"*: This Bahcall quote is from Johnson, "Elusive Particles Continue to Puzzle Theorists of the Sun," *The New York Times*, June 9, 1998.

101 *atmospheric neutrinos*: See Edward Kearns, Takaaki Kajita, and Yoji Totsuka, "Detecting Massive Neutrinos," *Scientific American*, August 1999, pp. 64–71.

101 *"That was the smoking gun"*: From an interview with Ed Kearns conducted by the author in person at Boston University on January 27, 2012.

102 *Sudbury Neutrino Observatory*: The official SNOLAB website is www .snolab.ca/. Other sources include an interview with Art McDonald conducted by the author via Skype on January 27, 2012; Nick Jelley, Arthur B. McDonald, and R.G. Hamish Robertson, "The Sudbury Neutrino Observatory," *Annual Review of Nuclear and Particle Science* 59 (2009): 431–65; and Arthur B. McDonald, Joshua R. Klein, and David L. Wark, "Solving the Solar Neutrino Problem," *Scientific American*, April 2003, pp. 40–49.

106 *"We've solved"*: This Art McDonald quote is from Kenneth Chang, "Sun's Missing Neutrinos: Hidden in Plain Sight," by *The New York Times*, June 19, 2001.

106 *"We now have high confidence"*: As quoted in a SNO press release issued on June 18, 2001 (www.sno.phy.queensu.ca/sno/first_results/).

107 *"Super-K told us"*: From the author's interview with Kearns.

107 *he felt like dancing*: As quoted in Chang, "Sun's Missing Neutrinos: Hidden in Plain Sight."

107 *"For three decades"*: This John Bahcall quote is from the *Nova* television documentary "The Ghost Particle," PBS, February 21, 2006. Transcript available at www.pbs.org/wgbh/nova/transcripts/3306_neutrino .html.

107 *"for pioneering contributions"*: From the Nobel Prize citation at www .nobelprize.org/nobel_prizes/physics/laureates/2002/. Ray Davis's share of the prize recognized his measurements of solar neutrinos, while Masatoshi Koshiba's share was primarily for the Kamiokande detection of neutrinos from Supernova 1987A (see chapter 6).

108 *"just a matter of time"* . . . *"It would be great"*: These Ed Kearns quotes are from an interview conducted by the author in person at Boston University on January 27, 2012. Other physicists who told me that Art McDonald deserves a share of a Nobel Prize include Janet Conrad, Karsten Heeger, and Scott Oser.

108 *"Traditional particle physics"*: From the author's telephone interview with Karsten Heeger, conducted on March 16, 2012.

109 *pitch, yaw, and roll*: This airplane analogy is borrowed from Dave Wark, as quoted by Jonathan Amos in "Neutrino Particle 'Flips to All Flavours,'" on the *BBC News* website, June 15, 2011; www.bbc.co.uk/news /science-environment-13763641.

110 *Soon after the discovery*: Here I am referring to the KamLAND experiment. See, for example, www.awa.tohoku.ac.jp/kamlande/ and http://arxiv .org/abs/hep-ex/0212021.

110 *MINOS experiment*: See the Fermilab press release of June 5, 2012 at https://www.fnal.gov/pub/presspass/press_releases/2012/minos-anti neutrinos-20120605.html.

111 *massive earthquake*: Martin Fackler, "Powerful Quake and Tsunami Devastate Northern Japan," *The New York Times*, March 12, 2011, provides a good overview of the devastation. The J-PARC project newsletter, May 2011, describes the damage at the Tokai site: http://j-parc .jp/hypermail/news-1.2011/0004.html. For a detailed account of the Fukushima nuclear plant disaster, see Eliza Strickland, "24 Hours at Fukushima," in *IEEE Spectrum*, November 2011, online at http://spectrum .ieee.org/energy/nuclear/24-hours-at-fukushima/0.

111 *Brian Kirby*: Kirby's quotes and account are from the interview conducted by the author via Skype on January 20, 2012.

111 *Scott Oser*: Oser's quotes and account are from the interview conducted in person by the author in Vancouver on January 15, 2012.

112 *data collected at T2K*: For a description of the T2K oscillation results, see http://legacy.kek.jp/intra-e/press/2011/J-PARC_T2Kneutrino.html and www.symmetrymagazine.org/breaking/2011/06/15/japans-t2k-experiment-observes-electron-neutrino-appearance.

113 *"Even though we have studied"*: Ed Kearns as quoted in a Boston University press release, June 15, 2011: www.bu.edu/phpbin/news/releases/display.php?id=2261.

113 *"the six most popular"*: This Lindley Winslow quote is from an interview conducted by the author in person at MIT on December 13, 2011.

113 *Double Chooz*: Cynthia Horwitz describes the experiment in "Power Plant Generates Neutrinos for Physics Experiment," *Symmetry*, February 23, 2011, online at www.symmetrymagazine.org/breaking/2011/02/23/power-plant-generates-neutrinos-for-physics-experiment, and the oscillation results are reported on the blog *Quantum Diaries*, November 9, 2011, at www.quantumdiaries.org/2011/11/09/first-double-chooz-neutrino-oscillation-result/ and www.interactions.org/cms/?pid=1031199.

113 *Daya Bay*: First oscillation results from the Daya Bay experiment are presented at http://neutrino.physics.berkeley.edu/news/News.html.

114 *RENO*: RENO's findings are described in a paper available at http://arxiv.org/abs/1204.0626 and in a press release at www.interactions.org/cms/?pid=1031612.

114 *"θ_{13} turns out"*: This and the other Kam-Biu Luk quotes are from a telephone interview conducted by the author on March 20, 2012.

114 *Janet Conrad*: The quotes "so incredibly beautiful" and "A detective is not always," and biographical information are from a telephone interview conducted by the author on March 4, 2013. The quote "We are entering" is from an interview conducted by the author in person at MIT on December 13, 2011.

6. EXPLODING STARS

120 *Supernova 1987A*: The story of the supernova discovery and aftermath is based on a number of sources including the *International Astronomical Union Circular No. 4316* issued by Brian G. Marsden (February 24, 1987; www.cbat.eps.harvard.edu/iauc/04300/04316.html); Ian K. Shelton,

"Supernova 1987A—Photometry of the Discovery and Pre-Discovery Plates," *Astronomical Journal* 105, no. 5 (1993): 1886–91, 2015–16; contemporary media accounts, in particular Michael D. Lemonick, Madeleine J. Nash, Gavin Scott, and Dick Thompson, "Supernova! Scientists Are Agog Over the Brightest Exploding Star in 383 Years," *Time*, March 23, 1987, p. 60; Nigel Henbest, "Supernova: The Cosmic Bonfire," in *New Scientist*, November 5, 1987, 52; and W. David Arnett, John N. Bahcall, Robert P. Kirshner, and Stanford E. Woosley, "Supernova 1987A," *Annual Review of Astronomy and Astrophysics* 27 (1989): 629–700.

121 *"It's like Christmas"*: Stan Woosley of the University of California at Santa Cruz, as quoted in *Time*, March 23, 1987.

121 *Bahcall and two of his colleagues*: Their short note is J. N. Bahcall et al., "Neutrinos from the Recent LMC Supernova," *Nature* 326 (March 12, 1987): 135–36.

121 *experimental physicists had begun*: Alfred K. Mann, *Shadow of a Star: The Neutrino Story of Supernova 1987A* (New York: W. H. Freeman, 1997) provides a detailed account of the detection of neutrinos from the supernova.

124 *"for the first time, we have"*: From Adam Burrows, "Supernova Neutrinos," *The Astrophysical Journal* 334 (November 15, 1988): 891–908.

124 *John Beacom*: All John Beacom quotes are from a telephone interview conducted by the author on March 19, 2012.

125 *"neutrino bomb"*: This and other Alex Friedland quotes are from a telephone interview conducted by the author on March 19, 2012.

125 *fiery saga of the star Sanduleak −69° 202*: For a good overview of stellar evolution, consult the widely used textbook by Bradley W. Carroll and Dale A. Ostlie, *An Introduction to Modern Astrophysics* (Boston: Addison-Wesley, 2006).

127 *"If a small fraction"*: This and other Georg Raffelt quotes are from a telephone interview conducted by the author on March 21, 2012.

128 *images of the site taken with the Hubble Space Telescope*: See, for example, http://hubblesite.org/newscenter/archive/releases/2010/30/full/.

130 *since 1604*: There is some evidence that Kepler's supernova of 1604 and Tycho's supernova of 1572 both resulted from the explosion of white dwarf stars, each of which gobbled up a companion star's material to reach the Chandrasekhar limit. If so, they would belong to the so-called Type Ia class of supernovae; see, e.g., http://chandra.harvard.edu/photo /2007/kepler/index.html, and http://physicsworld.com/cws/article/news

/2008/dec/03/echoes-shine-a-light-on-tycho-brahes-supernova. Supernova 1987A, on the other hand, is a Type II and marked the fiery death of a massive star.

130 *"a new and unusual star"*: Tycho's 1573 book *De Nova Stella* is available online at http://archive.org/details/operumprimitiasd00brahuoft. The English translation of the quote is from Robert Burnham, Jr., *Burnham's Celestial Handbook* (Mineola, NY: Dover Publications, 1978).

131 *"A new star of unusual . . . This spectacle appeared"*: As quoted by Laurence Marschall in *The Supernova Story* (Princeton: Princeton University Press, 1994), which also provides a superb popular account of other historical supernovae, how our understanding of supernovae evolved in the twentieth century, and the discovery and early studies of Supernova 1987A.

131 *Walter Baade*: See Donald E. Osterbrock: *Walter Baade: A Life in Astrophysics* (Princeton: Princeton University Press, 2001).

131 *Fritz Zwicky*: See Stephen Maurer, "Idea Man," in *Beam Line: A Periodical of Particle Physics*, Stanford Linear Accelerator Center, vol. 31, no. 1 (Winter 2001): 21–27, online at www.slac.stanford.edu/pubs/beam line/31/1/31-1-maurer.pdf.

133 *Kate Scholberg*: All Kate Scholberg quotes are from a telephone interview conducted by the author on March 16, 2012.

133 *SuperNova Early Warning System*: SNEWS maintains a website at http://snews.bnl.gov/. Additional useful sources include the popular article by Francis Reddy, "Time for SNEWS," at the *Astronomy* magazine website, www.astronomy.com/en/sitecore/content/Home/News-Observ ing/News/2005/06/Time%20for%20SNEWS.aspx; and the technical article by Pietro Antonioli, "SNEWS: The SuperNova Early Warning System," *New Journal of Physics* 6 (2004): 114.

136 *"We will be able"*: From the author's telephone interview with Francis Halzen, conducted on December 12, 2011.

136 *Helium and Lead Observatory*: Project website is at www.snolab.ca /halo/.

136 *Long-Baseline Neutrino Experiment*: See, e.g., Adrian Cho, "DOE Scraps Plans for Neutrino Experiment in Mine," *Science*Insider, March 22, 2012, http://news.sciencemag.org/scienceinsider/2012/03/doe-scraps -plans-for-neutrino.html; and Eugenie Samuel Reich, "US Physicists Fight to Save Neutrino Experiment," *Nature* News, March 26, 2012, www .nature.com/news/us-physicists-fight-to-save-neutrino-experiment-1 .10305.

138 *Laser Interferometer Gravitational-Wave Observatory*: By far the best popular account of the science and the technological challenge behind LIGO is Marcia Bartusiak, *Einstein's Unfinished Symphony: Listening to the Sounds of Space-Time* (Washington, D.C.: Joseph Henry Press/ National Academies Press, 2000).

140 *"diffuse supernova neutrino background"*: See John Beacom, "The Diffuse Supernova Neutrino Background," *Annual Review of Nuclear and Particle Science* 60 (November 2010): 439–62; preprint available online at http://arxiv.org/abs/1004.3311.

141 *Eta Carinae*: See, e.g., Michael Lemonick, "Supernova Countdown: Giant Star Could Explode Any Day Now," *Time*, February 16, 2012, www .time.com/time/health/article/0,8599,2106904,00.html; and the website of the Eta Carinae research group at the University of Minnesota, http:// etacar.umn.edu/.

7. VANISHING ACTS

143 *vast emptiness of the cosmos*: Parts of this chapter, including the quotes from Edward "Rocky" Kolb, first appeared in my article "Does Antimatter Matter?" in *Astronomy* magazine, December 2006, p. 30. A more extensive, semipopular-level account of the topic is Helen R. Quinn and Yossi Nir, *The Mystery of the Missing Antimatter* (Princeton: Princeton University Press, 2008). Rabindra N. Mohapatra's article "Neutrino Mass and the Origin of Matter," *Physics Today*, April 2010, p. 68, explains the connection between neutrinos and the matter-antimatter asymmetry.

145 *Paul Dirac*: The highly readable biography by Graham Farmelo, *The Strangest Man: The Hidden Life of Paul Dirac* (New York: Basic Books, 2009) served as a valuable source of information, anecdotes, and quotes on Dirac's life and work. Monica Dirac's article "My Father" in *Proceedings of the Dirac Centennial Symposium*, ed. Howard Baer and Alexander Belyaev (Singapore: World Scientific, 2003) provided insights into his family life. Several writers, including Victoria Brignell in the *New Statesman* ("How Autism Leads to Genius," online at www.newstatesman.com/society/2010 /11/dirac-autism-autistic), have suggested that Dirac was likely autistic.

148 *"an entirely new kind of matter"*: As quoted in Quinn and Nir, *The Mystery of the Missing Antimatter*.

148 *"something positively charged"*: Quoted from Carl Anderson's paper "The Apparent Existence of Easily Deflectable Positives," *Science* 76, no. 1967 (1932): 238–39.

149 *"As shy as a gazelle"*: From *The Sunday Dispatch* (1933), as quoted in Farmelo, *The Strangest Man*.

149 *antiproton*: Owen Chamberlain described the discovery of the antiproton in his 1959 Nobel Prize Lecture, www.nobelprize.org/nobel_prizes /physics/laureates/1959/chamberlain-lecture.pdf.

149 *antimatter versions of helium*: See the article by the STAR Collaboration, "Observations of the Anti-Matter Helium-4 Nucleus," *Nature* 473 (May 19, 2011): 353–56.

150 *Alpha Magnetic Spectrometer*: The experiment is described at http:// ams.nasa.gov.

151 *Emmy Noether*: Sources include Natalie Angier, "The Mighty Mathematician You've Never Heard Of," *The New York Times*, March 17, 2012; and Ransom Stephens, "The Unrecognized Genius of Emmy Noether," online at www.ransomstephens.com/emmy-noether.htm, accessed July 28, 2012.

152 *"one of the most important"*: From Leon M. Lederman and Christopher T. Hill, *Symmetry and the Beautiful Universe* (Amherst, NY: Prometheus Books, 2004).

153 *parity*: In 1956, T. D. Lee and C. N. Yang predicted that weak interactions might not conserve parity. The following year, C. S. Wu and her colleagues found evidence of parity violation in the beta decay of colbalt-60.

154 *Large Hadron Collider*: The LHCb experiment reported observations of CP violation in 2012: http://lanl.arxiv.org/abs/1202.6251.

155 *Ettore Majorana*: Sources of biographical information and quotes include Antonino Zichichi, "Ettore Majorana: Genius and Mystery," *CERN Courier*, July 25, 2006, http://cerncourier.com/cws/article/cern/29664; Barry Holstein, "The Mysterious Disappearance of Ettore Majorana," *Journal of Physics: Conference Series 173*, 012019 (2009); Salvatore Esposito, "Fleeting Genius," *Physics World*, August 2006, p. 2; and the biography by João Magueijo, *A Brilliant Darkness: The Extraordinary Life and Mysterious Disappearance of Ettore Majorana, the Troubled Genius of the Nuclear Age* (New York: Basic Books, 2009).

158 *"Majorana had greater gifts"*: Fermi is reported to have said this to Giuseppe Cocconi in the days following Majorana's disappearance, as quoted in, for example, *Ettore Majorana: Scientific Papers: On Occasion of the Centenary of His Birth*, ed. Giuseppe Franco Bassani and Council of the Italian Physical Society (Società Italiana di Fisica and Springer, 2006).

161 *"how difficult it is"*: This and other Giorgio Gratta quotes are from a telephone interview conducted by the author on March 19, 2012.

161 *CUORE*: Description of the use of lead from a sunken ship is based on Nicola Nosengo's report "Roman Ingots to Shield Particle Detector," *Nature* News, April 15, 2010, www.nature.com/news/2010/100415/full /news.2010.186.html; and Edwin Cartlidge's article "Ancient Romans Join Neutrino Hunt," *Physics World*, April 23, 2010, http://physicsworld .com/cws/article/news/2010/apr/23/ancient-romans-join-neutrino-hunt.

162 *EXO-200*: Sources include www.symmetrymagazine.org/article/febru ary-2010/exo-takes-clean-to-an-extreme and an interview with Giorgio Gratta.

163 *Hans Klapdor-Kleingrothaus*: Description of his results and interpretation is based on the author's interviews with Hitoshi Murayama, Giorgio Gratta, and Karsten Heeger, and on other sources including Edwin Cartlidge, "Double Trouble for Beta Decay," *Physics World*, July 2004, p. 8; see also March 2002, p. 5—online at http://physicsworld.com/cws/article /news/2002/feb/11/rare-decay-claim-stirs-controversy; and "Neutrino Physics: Beta Test," also by Cartlidge in *Nature*, July 12, 2012, pp. 160–62, online at www.nature.com/news/neutrino-physics-beta-test-1.10988.

164 *"Since it is such a profound"*: This and other Karsten Heeger quotes are from a telephone interview conducted by the author on March 16, 2012.

8. SEEDS OF A REVOLUTION

168 *Chasing the Higgs boson*: For accessible accounts, see Sean Carroll, *The Particle at the End of the Universe: How the Hunt for the Higgs Boson Leads Us to the Edge of a New World* (New York: Dutton, 2012); and Dennis Overbye, "Chasing the Higgs Boson," *The New York Times*, March 5, 2012, online at www.nytimes.com/2013/03/05/science/chas ing-the-higgs-boson-how-2-teams-of-rivals-at-CERN-searched-for -physics-most-elusive-particle.html; for historical background, see also Ray Jayawardhana, "Something for Nothing," in *Times Higher Education Supplement,* August 7, 1992, p. 15.

170 *"It's certainly been"*: Peter Higgs quotes are from a press conference at the University of Edinburgh, July 6, 2012; video available online at www .ed.ac.uk/news/all-news/120704-higgs July 6, 2012.

170 *"But it is a pity"*: Stephen Hawking quote is taken from Paul Rincon, "Higgs Boson-Like Particle Discovery Claimed at LHC," on the BBC News website, July 4, 2012, www.bbc.co.uk/news/world-18702455.

170 *firmed up the Higgs detection*: See the CERN press release "New Results Indicate That Particle Discovered at CERN is a Higgs Boson," March 14, 2013, online at http://press.web.cern.ch/press-releases/2013/03/new-results-indicate-particle-discovered-cern-higgs-boson.

170 *"Higgs is the last"*: All Steven Weinberg quotes in this chapter are from a telephone interview conducted by the author on August 10, 2012.

171 *"The zero mass"*: All Georg Raffelt quotes are from a telephone interview conducted by the author on March 21, 2012.

171 *"Zero we could"*: This André de Gouvêa quote is from a telephone interview conducted by the author on March 20, 2012.

172 *KATRIN*: See the project website at www.katrin.kit.edu.

174 *In fact, one of the best*: See, for example, Signe Riemer–Sørensen et al., "WiggleZ Dark Energy Survey: Cosmological Neutrino Mass Constraint from Blue High-Redshift Galaxies," *Physical Review D* 85, no. 8 (2012), online at http://arxiv.org/abs/1112.4940.

174 *Licia Verde*: Quotes from an e-mail interview conducted by the author on March 30, 2013, also informed this section.

174 *relic neutrinos*: See, for example, Andreas Ringwald, "Prospects for the direct detection of the cosmic neutrino background," http://arxiv.org/abs/0901.1529.

175 *Even geophysicists are*: Description of conflicting ages derived by Charles Darwin and William Thomson is based, in part, on "Fusion" at Nobelprize.org (2012), online at www.nobelprize.org/nobel_prizes/physics/articles/fusion/sun_1.html; also see Thomas Hayden, "What Darwin Didn't Know," *Smithsonian*, February 2009.

176 *"Emboldened by the remarkable"*: Lawrence Krauss quotes are from his article "Why I Love Neutrinos," *Scientific American*, June 2010.

177 *geoneutrinos*: Discussion of geoneutrinos and the Earth's internal heat is based on several sources, including Lawrence M. Krauss, Sheldon L. Glashow, and David N. Schramm, "Antineutrino Astronomy and Geophysics," *Nature* 310 (July 19, 1984): 191–98; Bertram M. Schwarzschild, "Neutrinos from Earth's Interior Measure the Planet's Radiogenic Heating," *Physics Today*, September 2011, p. 14; William McDonough, John Learned, and Stephen Dye, "The Many Uses of Electron Antineutrinos," *Physics Today*, March 2012, p. 46.

177 *Flash forward two decades*: See T. Araki et al., "Experimental Investigation of Geologically Produced Antineutrinos with KamLAND," *Nature* 436 (July 28, 2005): 499–503.

177 *"We now have a diagnostic"*: This Atsuto Suzuki quote is taken from

the Lawrence Berkeley National Laboratory news release titled "First Measurement of Geoneutrinos at KamLAND," July 27, 2005, online at http://newscenter.lbl.gov/news-releases/2005/07/27/first-measurement -of-geoneutrinos-at-kamland/.

177 *As of 2011*: See the KamLAND collaboration's article "Partial Radiogenic Heat Model for Earth Revealed by Geoneutrino Measurements," in *Nature Geoscience* 4 (2011): 647–51.

179 *to send neutrino beams*: See Steve Nadis, "'X-Raying' the Earth with Neutrinos," *Technology Review*, August 1997, online at www.technologyreview .com/article/400097/x-raying-the-earth-with-nutrinos/; C. A. Argüelles, M. Bustamante, and A. M. Gago, "Searching for Cavities of Various Densities in the Earth's Crust with a Low-Energy Electron-Antineutrino Beta-Beam," 2012, online at http://arxiv.org/abs/1201.6080.

180 *NEMO*: See Kathryn Jepsen, "Researchers Developing Underwater Neutrino Experiment Make Oceanographic Discovery," *Symmetry*, May 15, 2012, online at www.symmetrymagazine.org/breaking/2012/05/15 /researchers-developing-underwater-neutrino-experiment-make-ocean ographic-discovery/); Nicola Nosengo, "Underwater Acoustics: The Neutrino and the Whale," *Nature* 442 (2009): 560–61, online at www.nature .com/news/2009/091202/full/462560a.html.

182 *Their exploits could help*: See A. Bernstein, N. S. Bowden, A. Misner, and T. Palmer, "Monitoring the Thermal Power of Nuclear Reactors with a Prototype Cubic Meter Antineutrino Detector," in *Journal of Applied Physics* 103, 074905 (2008); J. R. Minkel, "To Catch a Plutonium Thief, Try Antineutrinos," *Scientific American* online, May 8, 2008, www .scientificamerican.com/article.cfm?id=to-catch-a-plutonium-thief-try -antineutrinos.

183 *"You need to bake"*: This and all other John Learned quotes are from a Skype interview conducted by the author on August 7, 2012.

185 *In a similar vein*: See Thierry Lasserre et al., "SNIF: A Futuristic Neutrino Probe for Undeclared Nuclear Fission Reactors," (2010), online at http://arxiv.org/abs/1011.3850.

186 *to send messages to submarines*: See Patrick Huber, "Submarine Neutrino Communication," *Physics Letters B* 692, no. 4 (2010): 268–71.

186 *They have used a neutrino beam*: See D. D. Stancil et al., "Demonstration of Communication Using Neutrinos," *Modern Physics Letters A* 27, 1250077 (2012).

187 *One researcher has suggested*: See Bruce Dorminey, "Neutrinos to Give

High-Frequency Traders the Millisecond Edge," *Forbes*, April 30, 2012, online at www.forbes.com/sites/brucedorminey/2012/04/30/neutrinos -to-give-high-frequency-traders-the-millisecond-edge/print/.

187 *to signal aliens*: John G. Learned, Sandip Pakvasa, and Anthony Zee, "Galactic Neutrino Communication," *Physics Letters B* 671, no. 1 (2009): 15–19.

187 *"All physics is wrong"*: This and other Boris Kayser quotes are from a telephone interview conducted by the author on August 9, 2012.

188 *Findings of NASA's*: See Gary Hinshaw et al., "Nine-Year Wilkinson Microwave Anisotropy Probe (WMAP) Observations: Cosmological Parameter Results," December 20, 2012, http://arxiv.org/abs/1212.5226; and Stephen M. Feeney, Hiranya V. Peiris, and Licia Verde, "Is There Evidence for Additional Neutrino Species from Cosmology?," last revised April 8, 2013, http://arxiv.org/abs/1302.0014.

188 *Planck spacecraft*: See the Planck collaboration's report "Planck 2013 Results. XVI. Cosmological Parameters," March 20, 2013, http://arxiv .org/abs/1303.5076.

188 *Janet Conrad*: Her quotes in this paragraph are from an e-mail interview conducted by the author on March 30, 2013.

189 *"High-energy collider"*: This Hitoshi Murayama quote is from a Skype interview with the author on March 28, 2012.

189 *Long-Baseline Neutrino Experiment*: For a description, see Kurt Riesselmann, "Long-Baseline Neutrino Experiment," *Symmetry*, February 2013, www.symmetrymagazine.org/article/february-2013/long-baseline -neutrino-experiment; and for a status update, see Toni Feder, "Dark Matter Search Gets Started Deep in Sanford Lab," *Physics Today*, February 2013.

190 *SNOLAB*: For an overview of the experiments to be hosted at SNOLAB, see, for example, Aksel Hallin and Doug Hallman, "The Wondrous New World of Modern Particle Astrophysics," *The Physics Teacher* 47 (May 2009): 274–80.

190 *Hyper-Kamiokande*: See K. Abe et al., "Letter of Intent: Hyper-Kamiokande—Detector Design and Physics Potential," September 15, 2011, http://arxiv.org/abs/1109.3262.

190 *LAGUNA*: See the project website at http://laguna.ethz.ch/LAGUNA /Welcome.html.

191 *"Whenever anything cool"*: This Lindley Winslow quote is from an interview conducted by the author in person at MIT on December 13, 2011.

ACKNOWLEDGMENTS

Writing often feels like a solitary pursuit, especially when tapping away on a laptop late at night while half the world sleeps, but this book would not have seen the light of day without the involvement of quite a few people. My late father's love of language rubbed off on me early, and a number of terrific editors have nurtured it over the years. For an active scientist, it is not always easy to find the time to write for a broad audience, but it can be rewarding and fun. Writing *Neutrino Hunters* gave me the chance to delve into a colorful and exciting field of science that is different from my own research area, and to appreciate its manifold connections to not only fundamental physics, cosmology, and astrophysics, but also geology and nuclear technology.

I am thankful to the many scientists who granted interviews, made helpful introductions, provided background material, and/or read parts of the book, including John Beacom, Sampa Bhadra, Janet Conrad, André de Gouvêa, Alex Friedland, Giorgio Gratta, Francis Halzen, Karsten Heeger, Christopher Jones, Boris Kayser, Ed Kearns, Brian Kirby, Spencer Klein, John Learned, Kam-Biu Luk, Art McDonald, Hitoshi Murayama, Scott Oser, Georg Raffelt, Kate Scholberg, Licia Verde, Steven Weinberg, and Lindley Winslow. Several of them not only tolerated repeated queries on my part but also responded promptly with their insights. Special thanks to Ralph Harvey for the ANSMET expedition, Nigel Smith for the SNO-LAB visit, Deborah Harris for the Fermilab tour, and Fred Raab for showing me the LIGO Hanford site. A number of people helped track down illustrations and secure permissions: Lili Bai of the Institute for Cosmic Ray Research in Japan, Erasmo Recami of Bergamo State University in Italy, Jane Koropsak of the Brookhaven National Laboratory, Samantha Kuula of SNOLAB, Laurel Norris of IceCube, and William Trischuk of the ATLAS collaboration.

It is a pleasure to acknowledge the wonderful colleagues and the stimulating environment (not to mention the fancy espresso machine!) at the Radcliffe Institute for Advanced Study at Harvard, where the book idea turned real. Natania Wolansky assisted with background research at the early stage, especially with tracking down pop culture references and gathering information about the many neutrino experiments around the world.

I am grateful to Amanda Moon at Scientific American / Farrar, Straus and Giroux for shepherding the manuscript with diligence, grace, and patience. Her editorial team, including Daniel Gerstle and Christopher Richards, and the production editor, Chris Peterson, truly excelled. So did the copy editor, Annie Gottlieb, who was not only attentive and thorough but also delightful. Laura Stephenson did a fine job with two of the illustrations. Jim Gifford at HarperCollins Canada and Robin Dennis at Oneworld UK provided enthusiastic support. My agent, John Pearce at Westwood Creative Artists, has been a constant source of encouragement, stepping in to smooth things over and cheer me on at difficult moments. Finally, I thank my family and friends, especially my awesome wife, Kathryn, for believing in me.

INDEX

Page numbers in *italics* refer to illustrations.

Printed in the USA
CPSIA information can be obtained
at www.ICGtesting.com
LVHW091141150724
785511LV00005B/461